Lecture Notes in Mathematics

Edited by A. Dold and B. Eckmann

652

Differential Topology, Foliations and Gelfand-Fuks Cohomology

Proceedings of the Symposium held at the Pontifíca Universidade Católica do Rio de Janeiro, 5–24 January, 1976

Edited by Paul A. Schweitzer, s.j.

Springer-Verlag
Berlin Heidelberg New York 1978

Editor
Paul A. Schweitzer
Departamento de Matemática
PUC/RJ
ZC-19
Rio de Janeiro
Brazil

Supported by FINEP with the collaboration
of CAPES, CNPq and FAPESP

Library of Congress Cataloging in Publication Data

Symposium on Differential and Algebraic Topology, Pontifíca Universidade Católica do Rio de Janeiro, 1976.
Differential topology, foliations, and Gelfand-Fuks cohomology.

(Lecture notes in mathematics ; 652)
Bibliography: p.
Includes index.
1. Differential topology--Congresses. 2. Algebraic topology--Congresses. 3. Homology theory--Congresses. I. Schweitzer, Paul A., 1937- II. Title. III. Series: Lecture notes in mathematics (Berlin) ; 652.
QA3.L28 no. 652 [QA613.6] 510'.8s
ISBN 0-387-07868-1 [514'.7] 78-8464

AMS Subject Classifications (1970): Primary: 57D20, 57D30, 22E65, 57E15
Secondary: 55B35, 57D45, 58D05, 55E15, 55F40, 54H25

ISBN 3-540-07868-1 Springer-Verlag Berlin Heidelberg New York
ISBN 0-387-07868-1 Springer-Verlag New York Heidelberg Berlin

Printing and binding: Beltz Offsetdruck, Hemsbach/Bergstr.
2141/3140-543210

Preface

This volume is essentially the Proceedings of the Symposium on Differential and Algebraic Topology (Escola de Topologia) held at the Pontifícia Universidade Católica do Rio de Janeiro from January 5 to 24, 1976. The central theme of the Symposium was the interaction of these two branches of topology, with special attention to foliations and the cohomology of the Lie algebra of vector fields on manifolds,following the pioneering work of I.M. Gelfand and D.B. Fuks.

A. Haefliger and R. Bott presented the two principal courses of lectures, on continuous cohomology, Gelfand-Fuks cohomology and characteristic classes of foliations. Aside from Haefliger's general introduction, "Cohomology of Lie algebras and foliations" (the first paper in this volume), only the new material from their courses is published here (in the second and third papers). References to earlier publications and an account of the full scientific program of the Symposium are given below.

A list of open problems, which has grown beyond its origins in the Symposium, appears at the end of the volume. The survey article by G. Reeb also suggests future lines of research on foliations. The papers of Langevin-Rosenberg, Reeb-Schweitzer, and Schachermayer were written and submitted subsequent to the Symposium, but the other papers represent lectures presented during the Symposium.

On behalf of the participants, I would like to extend heartfelt thanks to the many people who contributed to the success of the Symposium, especially my colleagues on the Organizing Committee, Gilberto Loibel, João Bosco Pitombeira de Carvalho, and William Whitley; the Symposium's excellent secretary, Isabel Viveiros de Castro Coffin; José Luis Arraut, former chairman of the Department of Mathematics, and many other personnel of PUC/RJ, too numerous to cite here. The advice, encouragement and assistance of colleagues from IMPA was indispensable. For the generous financial support which made the Symposium possible, we are grateful to Financiadora de Estudos e Projetos (FINEP) of the Brazilian government, the principal sponsor, and to the collaborating sponsors CAPES, CNPq, FAPESP, and PUC/RJ.

Paul A. Schweitzer,s.j.

CONTENTS

I. Gelfand-Fuks Theory and Characteristic Classes of Foliations

II. Qualitative Theory of Foliations

III. Group Actions and Related Topics

IV. Open Problems

DIFFERENTIAL TOPOLOGY, FOLIATIONS AND GELFAND-FUKS THEORY

PUC/RJ 1976

PROGRAM OF THE SYMPOSIUM

I. COURSES

R. Bott - CONTINUOUS COHOMOLOGY.

Ref. 1. R. Bott, M. Mostow and J. Perchik, Gelfand-Fuchs Cohomology and Foliations, Proceedings of the New Mexico State University Symposium, 1973 (mimeographed).

2. R. Bott, On some formulas for the characteristic classes of group-actions, these Proceedings.

A. Haefliger - ON THE GELFAND-FUKS COHOMOLOGY OF THE LIE ALGEBRA OF SMOOTH VECTOR FIELDS.

Ref. 1. A. Haefliger, Sur la cohomologie de l'algèbre de Lie des champs de vecteurs, Ann. Scient. École Norm.Sup., (4) 9 (1976),503-532.

2. ______, Sur la cohomologie de Gelfand-Fuchs, Differential Topology and Geometry (Dijon, 1974), Springer Lect. Notes Math.484 (1975), 121-161.

3. ______, Cohomology of Lie algebras and foliations, these Proceedings.

4. ______, Whitehead products and differential forms, these Proceedings.

S. Gitler - THE ADAMS SPECTRAL SEQUENCE.

II. SURVEY AND EXPOSITORY LECTURES

C. Camacho - Structural stability of foliations with singularities.*

J. Cheeger - The theory of differential characters (two lectures).

Ref. J. Cheeger and J. Simons, Differential Characters and geometric invariants (preprint - lecture notes, AMS Symposium, Stanford,1973).

D.B.A. Epstein - Foliations with all leaves compact.

Ref. D.B.A. Epstein, Foliations with all leaves compact, Ann. Inst. Fourier (Grenoble), 26(1976), 265-282.

* Published in these Proceedings

R. Gardner - Differential geometry and foliations: the Godbillon-Vey invariant and the Bott-Pasternack vanishing theorem.*

G. Reeb - Structures feuilletées.*

H. Rosenberg - The work of Arnold and Herman on Diff(S^1) (two lectures).
Ref. 1. H. Rosenberg, Les difféomorphismes du cercle (d'après M. R. Herman), Seminaire Bourbaki nº 476, Springer Lect. Notes Math. 567 (1977), 8-98.
2. M. Herman, Sur la conjugaison différentielle des difféomorphismes du cercle, thesis, Univ. de Paris (Orsay), 1977.

III. RESEARCH LECTURES

R. Barre - Q-varieties: methods to study transverse structures.
Ref. R. Barre, De quelques aspects de la théorie des Q-variétés différentielles et analytiques, Ann. Inst. Fourier (Grenoble) 23 (1973), 227-312.

F. Gonzalez Acuña - 3-dimensional open books.

A. Haefliger - Results on Hamiltonian vector fields.

G. Hector - Growth of foliations.*

F. Hegenbarth - The G-Signature Theorem and applications to surgery.*

J. Heitsch - Residues and characteristic classes for foliations.
Ref. 1. J. Heitsch, Independent variation of secondary classes (to appear).
2. ______, Residues, Γ-vector fields and foliations (to appear).

R. Knill - Stability of closed cycles and closed tori.*

A. Lins Neto - Local C^2- structural stability of integrable 1-forms.
Ref. A. Lins Neto, Local structural stability of C^2 integrable 1-forms, Ann. Inst. Fourier (Grenoble) 27 (1977), 197-225.

J. Palis - Diffeomorphisms and the stability of suspended foliations.*

M. Penna - Tangent bundles for polyhedra.
Ref. 1. M. Penna, Differential geometry on simplicial spaces, Trans. Amer. Math. Soc. 214 (1975), 303-323.

* Published in these Proceedings

2. _____, Vector fields on polyhedra, Trans. Amer. Math.Soc.232 (1977), 1-31.

B. Reinhart - Riemannian geometry and Godbillon-Vey classes.
Ref. B. Reinhart, The second fundamental form of a plane field, J. Differential Geometry (to appear).

P. Schweitzer - Pontryagin polynomial residues of isolated foliation singularities (joint work with A.Whitman).*

G. Segal - Thurston's theorem $BDiff_n^c \sim \Omega^n B\overline{\Gamma}_n$.
Ref. D. McDuff and G. Segal, A theorem of Thurston about the classifying space for foliations (to appear).

G. Segal - Proof of the Bott conjecture for Gelfand-Fuks chomology.
Ref. R. Bott and G. Segal, The cohomology of the vector fields on a manifold, Topology 16 (1977), 285-298.

H. Shulman - Flat bundles and the Van Est Isomorphism Theorem.
Ref. D. Tischler and H. Shulman, Leaf invariants of foliations and the Van Est isomorphism, J. Differential Geometry 11 (1976), 535 - 546.

A. Verjovsky - Minimal flows.

* Published in these Proceedings

PARTICIPANTS

Complete addresses for the institutes indicated in abbreviated form are given at the end of the list of participants.

Plácido F.deA. Andrade - PUC/RJ

José Luis Arraut - PUC/RJ

Luis Astey Q. - CIEA/IPN

Nélio Baldin - UNICAMP

Angelo Barone Netto - IME/USP

Guilhermo L. de la B. Alvarez
Departamento de Matemática
Universidad de Chile
Sede Valparaiso
Casilla 130
Valparaiso, CHILE

Raymond Barre
Département de Mathématique
Centre Universitaire de Valenciennes
59326 Valenciennes, FRANCE

Aristides C. Barreto - PUC/RJ

Raoul Bott
Department of Mathematics
Harvard University
One Oxford Street
Cambridge, MA 02138, USA

Cesar Camacho - IMPA

Alberto Campos S.
Departamento de Matemáticas
Universidad Nacional de Bogotá
Bogotá - COLOMBIA

Manuel Efrain Carbajal P.
Universidad Nacional Mayor de San Marcos
C.U. Pabellón D (Ciencias)
Av. Venezuela S/N
Lima, PERU

Carlos A. Aragão de Carvalho
Instituto de Matemática da UFRJ
Cidade Universitária
C.P. 1835 - ZC-00
20.000 - Rio de Janeiro, RJ
BRASIL

João B.Pitombeira de Carvalho
PUC/RJ

Jeff Cheeger
Department of Mathematics
SUNY at Stony Brook
Stony Brook, NY 11794, USA

Eduardo A. Chíncaro E.
Departamento de Matemática
Universidade Federal de Minas Gerais - Pampulha
30.000 Belo Horizonte, MG
BRASIL

Carlie Coats
Department of Mathematics
MIT - Room 2089
77 Massachusetts Ave.
Cambridge, MA 02139, USA

Sueli D.R. Costa - UNICAMP

Italo Dejter
Centro de Estudos Básicos
Universidade Federal de Santa Catarina - Departamento de Matemática - Conjunto Universitário Trindade
88.000 Florianópolis, SC
BRASIL

Freddy Dumortier
Institut de Mathématique
Université de Brussels
1050 Brussels, BELGIUM

David B.A. Epstein
Mathematics Institute
University of Warwick
Coventry CV4 7AL, UK

Luiz A. Fávaro - ICMSC

Samuel Feder- CIEA/IPN

Edmond Fédida
Département de Mathématique
Université de Dakar
Dakar, SENEGAL

Robert B. Gardner
Department of Mathematics
University of North Carolina
Chapel Hill, NC 27514, USA

Samuel Gitler - CIEA/IPN

Elza Gomide - IME/USP

Francisco Gonzãlez Acuña
Instituto de Matemáticas
Universidad Nacional Autónoma de México - Ciudad Universitária - México 20, D.F.-MÉXICO

André Haefliger
Institut de Mathématique
Université de Genève
C.P. 124
1211 Genève, SWITZERLAND

Gilbert Hector
UER de Mathématiques
Université de Lille I
B.P. 36
59650 Villeneuve d'Ascq,FRANCE

Abramo Hefez - IMPA

Friedrich Hegenbarth - UNICAMP
Universität Dortmund
Abteilung Mathematik
46 Dortmund, WEST GERMANY

James Heitsch
Department of Mathematics
Univ. of Illinois at Chicago
Circle - Box 4348
Chicago, IL 60680, USA

Wolf Iberkleid - CIEA/IPN

J. Carlos de S. Kiihl - UNICAMP

Ronald J. Knill
Department of Mathematics
Tulane University
New Orleans, LA 70118, USA

Maynard Kong
Departamento de Ciencias
Pontificia Universidade Católica
del Peru - Apartado 12514
Lima.21, PERU

Brasil Terra Leme
Departamento de Matemática
Universidade Federal de S. Carlos
13560 - S.Carlos, SP, BRASIL

Harold Levine
Department of Mathematics
Brandeis University
Waltham, MA 02154, USA

Alcides Lins Neto - IMPA

Gilberto Loibel - ICMSC

Artur O. Lopes
Instituto de Matemática
Universidade Federal do Rio
Grande do Sul
Rua Sarmento Leite S/N
90.000 - Porto Alegre, R.S.
BRASIL

Iaci P. Malta - IMPA

Solange Mancini
Faculdade de Filosofia e Letras
do Rio Claro
13.500 Rio Claro, S.P. - BRASIL

Ozíride Manzoli Neto - ICMSC

Welington de Melo - IMPA

Alan Mitchell - PUC/RJ

Jacob Palis - IMPA

C. Frederico Palmeira - PUC/RJ

Michael Penna
Department of Mathematics
Indiana University - Purdue
University at Indianapolis
1201 East 38th Street
Indianapolis, IN 46205, USA

Maria do Socorro O.Pereira
PUC/RJ

Harsh Pittie
Courant Institute of Mathematical Sciences
New York University
251 Mercer Street
New York, NY 10012, USA

Paulo F. da S. Porto Jr.-ICMSC

Georges Reeb
UER de Mathématiques
Université de Strasbourg
7, rue René Descartes
67084 Strasbourg Cedex, FRANCE

Bruce Reinhart
Department of Mathematics
University of Maryland
College Park, MD 20742, USA

Jean Roberts
Department of Mathematical
Sciences - Oakland Univ.
Rochester, MI 48063 - USA

Paulo C.P.Rodrigues - PUC/RJ

Harold Rosenberg
Departement de Mathématique
Univ. de París VII
2, place Jussieu
75005 Paris, FRANCE

Mario R. Saab - ICMSC

Cristián Sánchez
Instituto de Matemática, Astronomia y Física
Univ. Nac. de Córdoba
Córdoba 5000, ARGENTINA

Nathan M. dos Santos - PUC/RJ

Paul A. Schweitzer - PUC/RJ

Márcia Scialom - PUC/RJ

Graeme Segal
Mathematics Institute
St. Catherine's College
Oxford OXI 3UJ, UK

Herbert Shulman
Department of Mathematics
Belfer Graduate School of Science
Yeshiva Univ., 2495 Amsterdam Ave.
New York, NY 10033, USA

Hugo N. Torriani - UNICAMP

Ana Maria Urbina F.
Departamento de Matemáticas
Univ. de Chile - Sede Valparaiso
Casilla 130, Valparaiso, CHILE

Alberto Verjovsky - CIEA/IPN

William G. Whitley
Pós-Graduação em Matemática
Universidade Federal de Santa
Catarina
88.000 Florianópolis, SC. BRASIL

Andrew P. Whitman - PUC/RJ

H. Elmar Winkelnkemper
Department of Mathematics
College Park, MD 20742, USA

INSTITUTIONAL ADDRESSES

CIEA/IPN:
Departamento de Matemática
Centro de Investigación del I.P.N.
México 14, D.F., MÉXICO

ICMSC:
Instituto de Ciências Matemáticas de São Carlos
Av. Dr. Carlos Botelho, 1465
13.560 - São Carlos - S.P. - BRASIL

IME/USP:
Instituto de Matemática e Estatística
Universidade de São Paulo
C.P. 20.570 - Agência Iguatemi
01.451 - São Paulo, S.P. - BRASIL

IMPA:
Instituto de Matemática Pura e Aplicada
Rua Luiz de Camões, 68
20.000 - Rio de Janeiro, R.J.- BRASIL

PUC/RJ:
Departamento de Matemática
Pontifícia Universidade Católica do Rio de Janeiro
ZC - 19
20.000 - Rio de Janeiro, RJ, BRASIL

UNICAMP:
Instituto de Matemática
Universidade de Campinas
C.P. 1170
13.100 - Campinas - S.P.
BRASIL

COHOMOLOGY OF LIE ALGEBRAS AND FOLIATIONS

André HAEFLIGER

The aim of this first talk was to recall the link between the Gelfand-Fuchs cohomology of the Lie algebra of smooth vector fields on a manifold M and the theory of foliations. It served as a motivation for the computation of the Gelfand-Fuchs cohomology, because its elements appear as potential characteristic classes for foliated trivialized bundles with fiber M.

1. Gelfand-Fuchs cochains on the Lie algebra of vector fields.

Let M be a differentiable manifold. Denote by v_M the Lie algebra of smooth vector fields on M with the C^∞- topology (namely uniform convergence of any derivative on compact sets of M).

The Gelfand-Fuchs cochain algebra $C^*(v_M)$ on v_M (cf.[2]) will be the algebra of continuous multilinear alternating forms on v_M. An alternating k-form α on v_M is continuous if there exist continuous semi-norms $p_1,\dots,p_k$ on v_M such that, for any vector fields $v_1,\dots,v_k \in v_M$,

$$|\alpha(v_1,\dots,v_k)| \leq p_1(v_1)\dots p_k(v_k) . \tag{1}$$

The differential is defined by the formula

$$d\alpha(v_0,\dots,v_k) = \sum_{r<s} (-1)^{r+s} \alpha([v_r,v_s],\dots,\hat{v}_r,\dots,\hat{v}_s,\dots) . \tag{2}$$

The Jacobi identity implies that $d^2 = 0$. The cohomology of the differential algebra $C^*(v_M)$ will be denoted by $H^*(v_M)$ and called the cohomology of the topological algebra v_M.

The choice of the topology on v_M and of the differential will be fully justified by the proposition of § 3.

For a closed set A in M, define $v_A(M)$ to be the topological Lie algebra quotient of v_M by the ideal of those vector fields whose infinite jets vanish at each point of A. We shall also denote by $C^*(v_A(M))$ the differential algebra of the continuous alternating multilinear forms on $v_A(M)$ with the differential defined by (2). In particular, if 0 is a point in M, then $v_0(M)$ is isomorphic, via smooth local coordinates around 0, to the Lie algebra of formal vector fields on $R^n (n = \dim M)$. A continuous cochain on $v_0(M)$ depends only on finite order jets of vector fields at 0.

We shall denote by v_M^c the Lie algebra of smooth vector fields with compact support on M, considered as the direct limit over the compact sets K in M of the v_M^K, where v_M^K is the topological subalgebra of v_M of those vector fields with support in K. In that case, the elements of $C^*(v_M^c)$ are the alternating multilinear forms on v_M^c which are continuous when restricted to vector fields with support in a fixed compact set.

2. Foliated trivialized bundles with fiber M.

Let X be a differentiable manifold of dimension n. Let F be a smooth foliation on $X \times M$ transverse to the fibers of the projection of $X \times M$ on X. Namely F is a smooth completely integrable n-plane field $(x,y) \mapsto F(x,y) \subset T_{(x,y)}(X \times M)$ complementary to the tangent space of the fiber $\{x\} \times M$.

Let f be a smooth map of a differentiable manifold X' in X. The inverse image by the differential of $f \times id_M$ of the plane field F is a plane field $f^{-1}(F)$ which is completely integrable, and so a foliation on $X' \times M$ transverse to the fibers of the projection on X'.

Two foliations F_0 and F_1 on $X \times M$ transverse to the fibers of the projection on X are homotopic if there is a smooth foliation F on $[0,1] \times X \times M$ transverse to the fibers of the projection on $[0,1] \times X$ such that $i_k^{-1}(F) = F_k$, where i_k is the inclusion of X in $\{k\} \times M \subset [0,1] \times M$, $k = 0,1$.

A deformation of F_0 to F_1 is a smooth 1-parameter family F_t , $t \in [0,1]$, of foliations on $X \times M$ transverse to the fibers . If F_0 and F_1 are homotopic, then they are connected by a deformation, but not conversely in general.

A foliation F on $X \times M$ transverse to the fibers of the projection on X is with compact support (locally with respect to X) , if for each small enough open set U in X , the foliation restricted to $U \times M$ is horizontal (i.e. tangent to the slices $X \times \{y\}$) outside of the product of U with a compact set in M .

To give such a foliation F with compact support is equivalent to giving a smooth map h of the universal covering $\tilde{X}$ of X in the group Diff_M^c of diffeomorphisms of M with compact support (mapping the base point $\tilde{x}_0$ of $\tilde{X}$ to the identity) , and a representation φ of the fundamental group π of X in Diff_M^c such that

$$h(\gamma.\tilde{x}) = \varphi(\gamma)\ h(\tilde{x}) \quad \text{for each } \tilde{x} \in \tilde{X} \text{ and } \gamma \in \pi .$$

An homotopy between such foliations is equivalent to a smooth map of $\tilde{X} \times [0,1]$ in Diff_M^c (mapping $\tilde{x}_0 \times 0$ to the identity) compatible as above with a representation of π in Diff_M^c , while a deformation is equivalent to a smooth 1-parameter family of smooth maps of $\tilde{X}$ in Diff_M^c (mapping $\tilde{x}_0$ to the identity) compatible with a smooth family of representations φ_t of π in Diff_M^c .

For a closed set A in M , we shall also consider foliations on a neighbourhood of $X \times A$ in $X \times M$, transverse to the fibers of the projection on X ; two such foliations are identified if they coincide on a smaller neighbourhood of $X \times A$.

When A is a point 0 , then such foliations are just foliated trivialized microbundles.

3. <u>The characteristic homomorphism</u> (cf. [3]).

Let F be a smooth field of n-planes on $X \times M$ complementary to the field of tangent planes to the fibers $\{x\} \times M$.

Let v be a smooth vector field on X. There is a unique vector field $\tilde{v}$ on $X \times M$ contained in F (namely $\tilde{v}(x,y) \in F(x,y)$) and projecting on v. For any point $x \in X$, let $\theta(v(x))$ be the smooth vector field on M which is the projection on M of $\tilde{v}$ restricted to the fiber $\{x\} \times M$.

We can consider θ as an element of the vector space $A^1(X, v_M)$ of smooth 1-forms on X with value in v_M, i.e. as a smooth section of the topological vector bundle $TX^* \otimes v_M$, where TX^* is the dual of the tangent bundle of X. This 1-form θ will be called <u>the connection form associated to F</u>.

It is clear that conversely an element θ of $A^1(X, v_M)$ defines a smooth field F transverse to the fibers, namely $F(x,y)$ is the subspace of $T_{(x,y)} X \times M = T_x X \times T_y M$ spanned by the vectors $(v, \theta(v)(y))$, where v is any vector of $T_x X$.

For any 1-form $\theta \in A^1(X, v_M)$, we can define a smooth 2-form on X with values in v_M by

$$d\theta(v,w) = -\theta([v,w]) + v.\theta(w) - w.\theta(v)$$

where v and w are any smooth vector fields on X.

Here, $v.\theta(w)$ is the derivative in the direction of v of the family $x \mapsto \theta(w(x))$ of smooth vector fields on M depending differentiably on the parameter x.

We can also define the smooth 2-form $[\theta,\theta]$ on X with value in v_M by

$$[\theta,\theta](v,w) = [\theta(v),\theta(w)] - [\theta(w), \theta(v)]$$

$$= 2[\theta(v), \theta(w)] .$$

$A^*(X)$ will denote the differential algebra of smooth differential forms on X with the C^∞-topology. As in distribution theory, we put on $C^*(v_M)$ the topology of bounded convergence. With this topology the continuous dual of $C^1(v_M)$ is v_M (cf. [4]).

PROPOSITION. *The three following data are equivalent* :

1) *a smooth foliation* F *on* $X \times M$ *transverse to the fibers of the projection on* X

2) *a smooth 1-form* θ *on* X *with values in* v_M *such that*

$$d\theta + \tfrac{1}{2}[\theta,\theta] = 0$$

3) *a continuous morphism of differential graded algebras*

$$\chi : C^*(v_M) \to A^*(X) .$$

The form θ corresponding to F is called the *connection form* of F and the morphism $\chi_F : C^*(v_M) \to A^*(X)$ corresponding to F the *characteristic homomorphism* associated to F. This terminology is fully justified by the equivalence of 1) and 3).

Proof. We have already seen that there is equivalence between a smooth plane field F transversal to the fibers of the projection on X and a smooth 1-form $\theta \in A^1(X,v_M)$.

The field F is completely integrable if for any smooth vector fields v and w on X, we have

i) $$[\tilde{v},\tilde{w}] = \widetilde{[v,w]}$$

where $\tilde{v}$ denotes as before the lifting in F of v.

If we identify $\theta(v)$ with the vertical vector field on $X \times M$ whose value at (x,y) is $(0,\theta(v(x))(y)) \in T_xX \times T_yM$ and v with the horizontal vector field whose value at (x,y) is $(v(x),0)$, then we can write $\tilde{v} = v + \theta(v)$, and i) is equivalent to

ii) $$[v,\theta(w)] + [\theta(v),w] + [\theta(v),\theta(w)] = \theta([v,w])$$

As $[v,\theta(w)]$ is the vertical vector field $v.\theta(w)$, this last equation is equivalent to 2) .

Given F (or equivalently θ) , we can define the characteristic morphism χ_F by the formula

$$(\chi_F\alpha)(v_1,\ldots,v_k)(x) = \alpha(\theta(v_1(x)),\ldots,\theta(v_k(x)))$$

where α is a continuous k-linear alternating form on v_M and the v_i are vector fields on X . Because α is continuous, the form $\chi_F\alpha$ is of class C^∞ . This is a justification for using continuous forms on v_M .

Clearly χ_F is a continuous morphism of algebras. If condition 2) is satisfied, then it commutes with the differentials. This is a straightforward verification, using the classical formula

$$d\omega(v_1,\ldots,v_k) = -\sum_i(-1)^i v_i.\omega(v_1,\ldots,\hat{v}_i,\ldots,v_k)$$

$$+\sum_{r<s}(-1)^{r+s}([v_r,v_s],\ldots,\hat{v}_r,\ldots,\hat{v}_s,\ldots)$$

for the differential of a (k-1) - form on X .

This fact justifies the choice of the differential on $C^*(v_M)$.

Conversely, suppose that χ is a continuous morphism of algebras of $C^*(v_M)$ in $A^*(X)$. Fixing $x \in X$, we get a continuous linear map $C^1(v_M) \to T_x X^*$ by taking the restriction of $\chi(\alpha)$ to $T_x X^*$. As noted before, the continuous dual of $C^1(v_M)$ is v_M , so that the dual of this map will be a linear map $\theta_x : T_x X \to v_M$. Varying x , we get a smooth 1-form θ on X with values in v_M .

In fact, let $x_1,\ldots,x_n$ be local coordinates in X and $y_1,\ldots,y_m$ local coordinates in M . Let δ^i_y be the element of $C^1(v_M)$ associating to the vector field $\Sigma\, b_i\, \partial/\partial y_i$ the value of its component b_i in y . Then $\theta(v(x))$ is the smooth vector field $\Sigma < \chi(\delta^i_y),v(x)> \partial/\partial y_i$.

The morphism of $C^*(v_M)$ in $A^*(X)$ associated to θ is just χ, because this is true on $C^1(v_M)$ and by the fact that sums of products of 1-forms are dense in $C^*(v_M)$.

If χ commutes with d, then for any 1-form $\alpha \in C^1(v_M)$ and smooth vector fields v,w on X

$$-\alpha([\theta(v),\theta(w)] = v.\alpha(\theta(w)) - w.\alpha(\theta(v)) - \alpha\theta([v,w]) .$$

This implies condition 2 .

Remark. It is clear that the characteristic homomorphism is functorial, namely for a smooth map $f : X' \to X$, we have

$$f^* \circ \chi_F = \chi_{f^{-1}(F)} .$$

Equivalently, if for each $\alpha \in C^*(v_M)$, we denote $\chi_F(\alpha)$ by α_F, then

*) $$\alpha_{f^{-1}(F)} = f^*(\alpha_F) .$$

It is likely that the system of forms α_F is the only one verifying the functorial property *) .

Note also that for any non zero form $\alpha \in C^*(v_M)$, there is a foliation F on some $X \times M$ transverse to the fibers such that $\chi_F(\alpha) \neq 0$ (at least if M is compact) . Indeed, if $v_1,\dots,v_k$ are vector fields on M such that $\alpha(v_1,\dots,v_k) \neq 0$, let φ_t^i be the one parameter subgroup generated by v_i. Then for the foliation F on $R^k \times M$ defined by the map $h : R^k \to \mathrm{Diff}_M$

$$h(x_1,\dots,x_k) = \varphi_{x_1} \cdots \circ \varphi_{x_2} \cdots \circ \varphi_{x_k} ,$$

the form $\chi_F(\alpha)$ is non zero at the origin .

COROLLARY : The characteristic morphism χ_F induces a graded algebra morphism

$$H^*(\chi_F) : H^*(v_M) \to H^*(X,R)$$

functorial with respect to F and which depends only on the homotopy class of F. It varies smoothly under smooth deformations.

The images by $H(\chi_F)$ of elements of $H^*(v_M)$ are called smooth characteristic classes of F.

4. Generalization to other Lie algebras of vector fields.

Let $\mathfrak{g}$ be a topological Lie algebra of smooth vector fields on M. For instance, $\mathfrak{g}$ might be the Lie algebra of a Lie group acting effectively on M. Or $\mathfrak{g}$ could be the subalgebra of v_M of vector fields leaving invariant a volume or a symplectic form on M. Of course $\mathfrak{g}$ could be also the topological algebra v_M^c of vector fields with compact support on M.

A $\mathfrak{g}$-foliation F on $X \times M$, transverse to the fibers of the projection on X, is a smooth foliation as above such that the associated connection form θ is a smooth section of $A^1(X,\mathfrak{g})$, i.e. a smooth 1-form on X with values in $\mathfrak{g}$.

To each $\mathfrak{g}$-foliation F is associated a characteristic homomorphism of the differential algebra $C^*(\mathfrak{g})$ of continuous cochains in the algebra of smooth differential forms on X. Note that some care is needed for defining $C^*(\mathfrak{g})$; for instance in the case of v_M^c, the elements of $C^*(v_M^c)$ as defined in § 1 are multilinear forms which are only separately continuous.

With such precautions when $\mathfrak{g}$ is infinite dimensional, the proposition of § 3 is true with the same proof (provided $\mathfrak{g}$ is a reflexive space for the equivalence between 1) and 3); this is true in particular for v_M^c).

When A is a closed set in M and F a foliation on a neighbourhood of $X \times A$ transverse to the fibers of the projection on X, we still have a connection form $\theta \in A^1(X, v_A(M))$ and a characteristic homomorphism $\chi : C^*(v_A(M)) \to A^*(X)$. But in general it does not characterize uniquely the germ of F along $X \times A$, but only its infinite jet.

The consideration of the characteristic homomorphism is useful for extension problems of foliations.

Consider for instance a smooth foliation F_A on a neighbourhood of $X \times A$, transverse to the fibers of the projection on X. Suppose that α is a cocycle in $C^*(v_A(M))$ whose image by the morphism $C^*(v_A(M)) \to C^*(v_M)$ vanishes, but whose image by the characteristic homomorphism is a non exact form. Then it is clear that it is impossible to construct a foliation F on $X \times M$ transverse to the fibers which agrees with F_A on A.

The analogous considerations apply when we want to construct a foliation on $X \times M$, transverse to the fibers, which extends a foliation given on $Y \times M$, where Y is a subspace of X.

5. The classifying space $B\mathfrak{g}$ and the differentiable cohomology.

Suppose now that X is a simplicial complex. A $\mathfrak{g}$-foliation F on $X \times M$ transverse to the fibers of the projection on X is, for each simplex σ of X, a $\mathfrak{g}$-foliation $F(\sigma)$ transverse to the fibers, such that the restritions above the intersection of two simplices σ and τ of $F(\tau)$ coincide.

For such a foliation, we also have a characteristic homomorphism $\chi : C^*(\mathfrak{g}) \to A^*(X)$, where $A^*(X)$ denotes the differential algebra of differentiable forms on X in the sense of Sullivan (namely a form ω on X is a smooth n-form $\omega(\sigma)$ for each simplex σ of X, such that the restrictions of $\omega(\mathfrak{g})$ and $\omega(\tau)$ on $\sigma \cap \tau$ coincide).

More generally, we can take for X a simplicial set. A $\mathfrak{g}$-foliation on $X \times M$ transverse to the projection on X is, for each k-simplex σ of X, a $\mathfrak{g}$-foliation $F(\sigma)$ on $\Delta^k \times M$ transverse to the fibers, such that for each face or degeneracy operator h^* induced by an affine map : $\Delta^\ell \to \Delta^k$, we have $F(h^*\sigma) = h^{-1}F(\sigma)$.

Similary a differential form ω on X is, for each k-simplex σ, a smooth differential form $\omega(\sigma)$, such that $\omega(h\sigma) = h^*\omega(\sigma)$. On the differential algebra $A^*(X)$ of smooth forms on X, we put the topology of uniform convergence of all derivatives on each simplex of X.

There is always a continuous characteristic homomorphism

$$\chi : C^*(\mathfrak{g}) \to A^*(X)$$

and proposition 3 is valid (provided it is true for $\mathfrak{g}$-foliations on $\Delta^k \times M$). Recall that the cohomology of $A^*(X)$ is the real cohomology of X (cf. Sullivan [6]).

For $\mathfrak{g}$-foliations transverse to the fibers M over a simplicial set, there is a classifying simplicial set $B\mathfrak{g}$. Its k-simplexes are $\mathfrak{g}$-foliations on $\Delta^k \times M$ transverse to the fibers of the projection on Δ^k; the boundary and degeneracy operators are obtained as restrictions or pull backs in an obvious way.

When proposition 3 is valid, the set of k-simplexes of $B\mathfrak{g}$ is the same as the set of continuous morphisms of $C^*(\mathfrak{g})$ in $A^*(\Delta^k)$ (cf. [6], § R). So it depends only on $\mathfrak{g}$: this justifies the notation $B\mathfrak{g}$ (due to G. Segal). When $\mathfrak{g}$ is the Lie algebra v_M^c or the Lie algebra of a group G acting on M, then $B\mathfrak{g}$ is also denoted $B\overline{\mathrm{Diff}}{}_M^c$ or $B\overline{G}$ (notation of Thurston [4]). When $\mathfrak{g}$ is the Lie algebra $v_0(R^n)$ of formal vector fields on R^n, then $B\mathfrak{g}$ is ususally denoted by $B\overline{\Gamma}_n$ or $F\Gamma_n$.

A differential form ω on $B\mathfrak{g}$ is smooth under deformation if, for any smooth family σ_t of k-simplexes of B depending on a parameter t in R^n (i.e. a differentiable family of $\mathfrak{g}$-foliations on $\Delta^k \times M$), then $\omega(\sigma_t)$ is a smooth family of forms on Δ^k.

The differential forms on $B\mathfrak{g}$ smooth under deformations form a differential subalgebra $A_d^*(B\mathfrak{g})$ of $A^*(B\mathfrak{g})$. Its cohomology could be called the <u>differentiable cohomology of</u> $B\mathfrak{g}$ (cf. [1]), and will be denoted by $H_d^*(B\mathfrak{g})$.

On the other hand, the cohomology of $A^*(B\mathfrak{g})$ is the real cohomology of $B\mathfrak{g}$. The inclusion induces a morphism

$$H_d^*(B\mathfrak{g}) \to H^*(B\mathfrak{g}) .$$

The universal characteristic morphism χ maps $C^*(\mathfrak{g})$ injectively (cf. remark of § 3) in the subalgebra $A^*_d(B\mathfrak{g})$.

$$\begin{array}{ccc} C^*(\mathfrak{g}) & \xrightarrow{\ \chi\ } & A^*(B\mathfrak{g}) \\ & \searrow \quad \nearrow & \\ & A^*_d(B\mathfrak{g}) & \end{array}$$

In many cases, the map $C^*(\mathfrak{g}) \to A^*_d(B\mathfrak{g})$ induces an isomorphism in cohomology. In the case of the Lie algebra of a Lie group, this is essentially one of the forms of the Van Est theorem.

The real cohomology of $B\mathfrak{g}$ (which is the important one for the extension problems of foliations, cf. § 4) is very hard to compute ; even in the simplest cases, it is very complicated. For instance for the Lie algebra $\mathfrak{r}$ of the additive group $\mathbb{R}$, $B\mathfrak{r}$ is the classifying space for the group $\mathbb{R}$ with the discrete topology ; its cohomology is the algebra of alternating forms on $\mathbb{R}$ with values in $\mathbb{R}$ which are multilinear over the rationals $\mathbb{Q}$.

On the other hand, the cohomology of the Lie algebra is much simpler and easier to compute. So the hope is that the properties of the cohomology of $\mathfrak{g}$ will reflect some properties of the cohomology of $B\mathfrak{g}$.

Among the main problems, let us mention once more

a) To what extent is the characteristic morphism $H^*(\mathfrak{g}) \to H^*(B\mathfrak{g})$ injective ?

b) What classes of $H^*(\mathfrak{g})$ can vary continuously ; more precisely, given an element of $H^*(\mathfrak{g})$, what values can its image under the characteristic homomorphism evaluated on the integral homology of $B\mathfrak{g}$ take ?

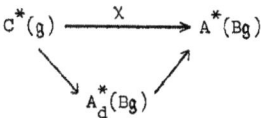

R E F E R E N C E S

[1] R. BOTT Some Remarks on Continuous Cohomology, Conference on Manifolds, Tokio 1973.

[2] GELFAND-FUCHS : The cohomology of the Lie algebra of tangent vector fields on a smooth manifold I, Functional Analysis, Vol. 3 (1969) , p. 32-52 .

[3] A. HAEFLIGER : Sur les classes caractéristiques des feuilletages. Séminaire Bourbaki (1970-71) , N° 412 .

[4] W. THURSTON : Foliations and groups of diffeomorphisms. Bull. Amer. Math. Soc. 80 (1974) , p. 304-307 .

[5] L. SCHWARTZ : Théorie des distributions, Hermann, 1957 .

[6] D. SULLIVAN : Infinitesimal Computations in Topology (To appear) .

WHITEHEAD PRODUCTS AND DIFFERENTIAL FORMS

André HAEFLIGER

This talk served as an introduction for the theorem asserting that the cohomology of the Lie algebra of vector fields on R^n with compact supports is isomorphic to the real cohomology of the n-th loop space of a wedge of spheres (cf.[2]).

1. Definition of Whitehead products.

Let D^n be the set of vectors in R^n of length ≤ 1 and $\partial D^n = S^{n-1}$ its boundary. Let X be a topological space with a base point x_o.

Given two continuous maps $f_i : (D^{p_i}, \partial D^{p_i}) \to (X, x_o)$ $i = 1,2$, the Whitehead product of f_1 and f_2 is the map $[f_1, f_2]$ of $\partial(D^{p_1} \times D^{p_2})$ in X defined by :

$$[f_1, f_2](x_1, x_2) = \begin{cases} f_1(x_1) & \text{for } x_2 \in \partial D^{p_2} \\ f_2(x_2) & \text{for } x_1 \in \partial D^{p_1} . \end{cases}$$

The f_i represent elements φ_i of the homotopy groups $\pi_{p_i}(X, x_o)$. The homotopy class $[\varphi_1, \varphi_2] \in \pi_{p_1+p_2-1}(X, x_o)$ of $[f_1, f_2]$ depends only on φ_1 and φ_2 and is called the Whitehead product of φ_1 and φ_2.

With the correct grading and the correct sign in front of the Whitehead product, the homotopy groups of X tensored by the reals R form a graded Lie algebra over R.

A graded Lie algebra L is a collection of vector spaces L_q, q an integer, with a bilinear map

$$L_p \times L_q \xrightarrow{[,]} L_{p+q}$$

such that

$$[x,y] = -(-1)^{\deg x \deg y}[y,x]$$

$$[x,[y,z]] = [[x,y],z] + (-1)^{\deg x \deg y}[y,[x,z]] .$$

For instance one gets such a graded Lie algebra over R by considering $\pi_*(X)\otimes R$ as a graded vector space whose homogeneous component of degree q is $\pi_{q+1}(X)\otimes R$, and defining the bracket of $\varphi_1\otimes 1\in\pi_{p_1}(X)\otimes R$ and $\varphi_2\otimes 1\in\pi_{p_2}(X)\otimes R$ as $(-1)^{p_1}[\varphi_1,\varphi_2]\otimes 1$.

2. Using forms to detect Whitehead products.

Recall that when X is a differentiable manifold, one can use differentiable maps and homotopy to define the homotopy groups.

We begin with a lemma often used for detecting the Hopf invariant.

LEMMA 1. *Let ω_1 and ω_2 be two forms on the differentiable manifold X of degrees p_1 and p_2 greater than one. Assume that*

$$d\omega_1 = d\omega_2 = 0 \quad \text{and} \quad \omega_1\omega_2 = 0 .$$

For a smooth map $f : S^{p_1+p_2-1}\to X$, the form $f^*\omega_1$ is of degree $p_1 < p_1+p_2-1$, so there is a form α_1 such that $d\alpha_1 = f^*\omega_1$.

The number

$$h_f = h_f(\omega_1,\omega_2) = \int_{S^{p_1+p_2-1}} \alpha_1 . f^*\omega_2$$

is independent of the choice of α_1 and of f in its homotopy class. It defines a homomorphism of $\pi_{p_1+p_2-1}(X)$ in R.

COROLLARY. If $h_f(\omega_1,\omega_2)\neq 0$, then f represents an element of infinite order in $\pi_{p_1+p_2-1}(X)$.

Proof. Suppose α_1' is another form such that $d\alpha_1' = f^*\omega_1$. There is a (p_1-2)-form β on $S = S^{p_1+p_2-1}$ such that $d\beta = \alpha_1 - \alpha_1'$. Then by Stokes' formula :

$$\int_S \alpha_1 . f^*\omega_2 - \int_S \alpha_1' . f^*\omega_2 = \int_S d(\beta . f^*\omega_2) = 0 .$$

Let F be a smooth homotopy of $S \times I \to X$, where I is the interval $[0,1]$, connecting f_o to f_1. Let α_1 be a form on $S \times I$ such that $F^*\omega_1 = d\alpha_1$. Let i_k be the inclusion $x \mapsto (x,k)$ of S in $S \times I$.

Then $h_{f_k}(\omega_1,\omega_2) = \int_S i_k^*(\alpha_1 \wedge F^*\omega_2)$,

hence

$$h_{f_1}(\omega_1,\omega_2) - h_{f_o}(\omega_1,\omega_2) = \int_{S\times\partial I} \alpha_1 \wedge F^*\omega_2 = \int_{S\times I} d\alpha_1 \wedge F^*\omega_2$$

$$= \int_S F^*(\omega_1\omega_2) = 0 .$$

THEOREM. *Let ω_1 and ω_2 be two differential forms on X such that $d\omega_1 = d\omega_2 = 0$ and $\omega_1\omega_2 = 0$.*

Let $f_i : (D^{p_i}, \partial D^{p_i}) \to (X, x_o)$ be smooth maps.

Then for $f = [f_1, f_2]$,

$$h_f(\omega_1,\omega_2) = \omega_1(f_1)\omega_2(f_2) + (-1)^{p_1p_2}\omega_1(f_2)\omega_2(f_1)$$

where
$$\omega_i(f_j) = \int_{D^{p_j}} f_j^*\omega_i .$$

Proof. Let $\pi_i : D^{p_1} \times D^{p_2} \to D^{p_i}$ be the natural projections $(i = 0,1)$. Choose forms β_i on D^{p_1} and γ_i on D^{p_2} such that $f_1^*\omega_i = d\beta_i$ and $f_2^*\omega_i = d\gamma_i$.

Then, if j is the inclusion $\partial(D^{p_1} \times D^{p_2}) \to (D^{p_1} \times D^{p_2})$,

$$f^*(\omega_i) = j^*(\pi_1^*f_1^*\omega_i + \pi_2^*f_2^*\omega_i) = dj^*(\pi_1^*\beta_i + \pi_2^*\gamma_i) .$$

Let $\alpha_1 = j^*(\pi_1^*\beta_1 + \pi_2^*\gamma_1)$; so $d\alpha_1 = f^*\omega_1$.

Hence by Stokes' formula :

$$h_f(\omega_1,\omega_2) = \int_{\partial(D^{p_1}\times D^{p_2})} \alpha_1 . f^*\omega_2 =$$

$$\int_{D^{p_1}\times D^{p_2}} (\pi_1^* f_1^*\omega_1 + \pi_2^* f_2^*\omega_1)(\pi_1^* f_1^*\omega_2 + \pi_2^* f_2^*\omega_2)$$

$$= \int_{D^{p_1}} f_1^*\omega_1 \int_{D^{p_2}} f_2^*\omega_2 + (-1)^{p_1p_2} \int_{D^{p_2}} f_2^*\omega_1 \int_{D^{p_1}} f_1^*\omega_2 \ .$$

Applications.

1. Let $X = S^n$ and let ω be a n-form on S^n such that $\int_{S^n} \omega = 1$. Let f be a smooth map of degree one of D^n on S^n mapping ∂D^n to a point. Then

$$h_{[f,f]}(\omega,\omega) = \begin{cases} 0 & n \text{ odd} \\ 2 & n \text{ even} \end{cases} .$$

This follows from the theorem because $\int_{D^n} f^*\omega = \int_{S^n} \omega = 1$.

2. Let X be the complement of $(1,0,0)$ and $(-1,0,0)$ in R^3 . Then X retracts by deformation on the union of the 2-spheres S_+ and S_- of radius 1 and center $(1,0,0)$ and $(-1,0,0)$. Let ω_+' be a 2-form on S_+ whose support is a small neighbourhood of $(2,0,0)$ and such that $\int_{S_+} \omega_+' = 1$. Let ω_+ be the 2-form $\rho_+^*\omega_+'$, where ρ_+ is the radial retraction of X on S_+ . Then ω_+ is a closed 2-form whose support is contained in the half-plane $x_1 > 0$. Let ω_- be the image of ω_+ by the symmetry with respect to the plane $x_1 = 0$. It is clear that $\omega_+ \omega_- = 0$.

Let i_+ and i_- be the two natural inclusions of S^2 in X with image S_+ and S_-. We want to show that the element $[i_+,i_-]$ of $\pi_3(X) = \pi_3(S^2 \vee S^2)$ is non zero.

As $\int_{S_+} \omega_+ = \int_{S_-} \omega_- = 1$ and $\int_{S_+} \omega_- = 0$, we have

$$h_{[i_+,i_-]}(\omega_+,\omega_-) = 1 .$$

It is clear that a similar argument can be used for the wedge of two spheres S^{p_1} and S^{p_2} , $p_i > 1$.

Remark. As it was pointed out to me by my auditors, in particular S. Gitler and H. Shulman, in the last theorem it is not necessary to assume that $\omega_1 \omega_2 = 0$ but only that $\omega_1 \cdot \omega_2 = d\beta$.
Indeed, if we redefine $h_f(\omega_1,\omega_2)$ as $\int_S (\alpha_1 f^*\omega_2 - f^*\beta)$ and if we assume that $f^*H^{p_1+p_2-1}(X) = 0$, then $h_f(\omega_1,\omega_2)$ is well defined and is the functional cup product of Steenrod (cf.[6] p.987).

When $f = [f_1,f_2]$, the extra hypothesis is verified, because f can be **factored** through a map in $S^{p_1} \vee S^{p_2}$.

A similar remark should apply to what follows.

Generalization to iterated Whitehead products.

LEMMA 1'. Let $\omega_1,\omega_2,\omega_3$ be three closed forms on X such that $\omega_1\omega_3 = 0$ and $\omega_2 \omega_3 = 0$. Let p_i = degree of $\omega_i > 1$. Let f be a smooth map of $S = S^{p_1+p_2+p_3-2}$ in X .

Choose forms α_1 and α_2 on S such that $d\alpha_i = f^*\omega_i$.

Then the real number

$$h_f(\omega_1,\omega_2,\omega_3) = \int_S \alpha_1 \cdot \alpha_2 \cdot f^*\omega_3$$

is independent of the choices of α_1,α_2 and f in its homotopy class. It defines a homomorphism of $\pi_{p_1+p_2+p_3-2}(X)$ in R .

THEOREM '. For $i = 1,2,3$, let ω_i be a closed form of degree $q_i > 1$, such that $\omega_i \omega_j = 0$ for $i \neq j$.

Let $f_i : (D^{p_i}, \partial D^{p_i}) \to (X, x_o)$ be smooth maps and assume $q_1+q_2+q_3 = p_1+p_2+p_3$. Then for $f = [f_1,[f_2,f_3]]$

$$h_f(\omega_1,\omega_2,\omega_3) = \omega_1(s_1)\omega_2(s_2)\omega_3(s_3) + (-1)^{p_2p_3}\omega_1(s_1)\omega_2(s_3)\omega_3(s_2)$$
$$+(-1)^{(p_1-1)(p_2-1)}\omega_2(s_1)[\omega_1(s_2)\omega_3(s_3) + (-1)^{p_1p_3}\omega_1(s_3)\omega_3(s_2)] ,$$

where $\omega_i(s_j) = \int_{D^{p_j}} f_j^*(\omega_i)$.

The proof is left to the reader. It is clear that one could continue in this way for higher Whitehead products, if one is not afraid of complicated formulas.

We can apply this theorem to the example 2) above, taking $\omega_1 = \omega_3 = \omega_+$ and $\omega_2 = \omega_-$. One can check that $[i_+,[i_+,i_-]] \neq 0$. Similarly, one can check that $[i_-,[i_+,i_-]]$ is linearly independent of the previous element. More generally, using just those two forms ω_+ and ω_-, one could show that for $S^2 \vee S^2$, there is an infinite number of non zero homotopy groups.

The next approach is less elementary, but more powerful.

3. Hilton's theorem and Sullivan theory.

Let p be a sequence $(p_1,\ldots,p_k)$ of integers >1. Denote by $\vee S^p$ the wedge of the spheres $S^{p_1},\ldots,S^{p_k}$.

Let V be the graded vector space over R with a basis $x_1,\ldots,x_k$ with $\deg x_i = p_i-1$.

The theorem of Hilton asserts that the graded Lie algebra $\pi_*(\vee S^p)\otimes R$ described in § 1 is the free graded Lie algebra $L(V)$ on V (cf.[3]).

An ordered basis of $L(V)$ whose elements are iterated brackets of x_i can be constructed inductively (see for instance Bourbaki [1] for the non graded case) ; one also has a formula (cf.[1]) counting the number of elements of this basis made up with brackets of k elements. This number increases very rapidly with k. For instance, the rank of $\pi_{34}(S^2 \vee S^2)\otimes R$ is 260,300,986.

In the rest of this paragraph, we want to sketch the construction of a minimal model in the sense of Sullivan for the cohomology algebra $H^*(\vee S^p;R)$. This plays a central role in the computation of Gelfand-Fuchs cohomology (cf.[2]). It also gives a way of describing the rational homotopy groups of $\vee S^p$.

Brief account of Sullivan theory (cf.[7]).

We shall work in the category of positively graded differential algebras A which are commutative in the graded sense : $a\,a' = (-1)^{\deg a \deg a'} a'\,a$.

A 1-connected minimal algebra M is the free symmetric algebra $S_*(V)$ over a graded vector space V with $V^q = 0$ for $q \leq 1$. (Symmetric means polynomial in even dimensional generators, and exterior in the odd ones.) The differential d of any generator has to be decomposable ; in other words, $dM^+ \subset M^+.M^+$, where M^+ are the elements of degree >0.

For any 1-connected algebra A (i.e. $H^0(A) = H^1(A) = 0$), there is a minimal model, namely a minimal algebra M together with a morphism $\mu : M \to A$ inducing an isomorphism in cohomology. The existence of M is easy to prove, but sometimes its explicit description might be very complicated.

One of the important links of the minimal model with geometry is the following property. Let $A^*(X)$ be the differential algebra of forms on a manifold X which is 1-connected and with finite dimensional cohomology. If M is a minimal model of $A^*(X)$, then the vector space $M^+/M^+.M^+$ of its indecomposable elements is isomorphic to the dual of $\pi_*(X) \otimes R$.

Cohomology of a graded Lie algebra.

If L is a Lie algebra over R in the usual sense, then $C^*(L)$ is the differential algebra of multilinear alternating forms on L with the differential defined as in the first talk [9].

For a graded Lie algebra $L = \{L_q\}$, one can construct an analogous algebra which generalizes the usual case (where L is considered as a graded Lie algebra with only one nontrivial component in degree 0). One first considers the graded vector space ΣL whose component of degree q is L_{q-1}. Then $C^*(L)$

is the algebra of multilinear forms on ΣL, symmetric in the graded sense (namely symmetric in even dimensional variables and antisymmetric in the odd ones). A differential is defined in $C^*(L)$ in terms of the bracket by the same formula as usual, but with the correct signs (cf.[2] and [5]). Note that $C^*(L)$ is a 1-connected minimal model if $L_q = 0$ for $q \leq 0$.

The cohomology of this algebra is the cohomology of the graded Lie algebra L. It can also be defined in terms of projective resolutions of R over the envelopping algebra UL of L (cf.[4] and [5]).

For a free graded Lie algebra $L = L(V)$ over the graded vector space V, the short exact sequence

$$0 \to \mathrm{Ker}\,\varepsilon \to UL \xrightarrow{\varepsilon} R \to 0$$

where ε is the augmentation, is a free UL-resolution, because $\mathrm{Ker}\,\varepsilon$ is isomorphic as a UL-module to $UL \otimes V$ (cf. for instance [4], p. 232 for the non graded case). Hence the cohomology of $C^*(L)$ is the same as the cohomology of the dual over UL of the complex $0 \to \mathrm{Ker}\,\varepsilon \to U(L) \to 0$.

It follows that the reduced cohomology H^+ of $C^*(LV)$ is isomorphic to the dual of ΣV, and has a trivial multiplicative structure.

Moreover let μ be the map of $C^*(L(V)) \to H^*(L(V))$ sending a k-linear form to zero if $k > 1$, and for $k = 1$ to its restriction to $\Sigma V \subset \Sigma LV$. Then μ is a differential algebra morphism inducing an isomorphism in cohomology. So we get

THEOREM. <u>The minimal model for the algebra $H^*(\vee S^P)$ (i.e. for a 1-connected algebra with trivial differential and trivial multiplication) is the differential algebra $C^*(L(V))$ of cochains on the free Lie algebra $L(V)$ over the graded vector space V, where V_q is the dual of $H^{q+1}(\vee S^P)$ and $V_{-1} = 0$.</u>

Note that there is a differential algebra morphism i of $H^*(\vee S^P)$ into $A^*(\vee S^P)$ inducing an isomorphism in cohomology. It is sufficient for that to map the natural generator corresponding to S^{P_r} to a form with support in

$S^{p_r} \subset \vee S^p$ whose integral over S^{p_r} is 1 .

So $i \circ \mu$ is also a minimal model of $A^*(\vee S^p)$. Hilton's theorem is then a particular case of the result of Sullivan mentioned above, because the vector space of indecomposable elements of $C^*(L(V))$ is isomorphic to the dual of $\Sigma L(V)$.

4. A theorem.

Let $\vee S$ be the wedge $S^{p_1} \vee \dots \vee S^{p_k}$, where the p_k are integers > 1 . Let i_r be the inclusion of S^{p_r} in $\vee S$.

THEOREM. *Let $\omega_1, \dots, \omega_k$ be closed differential forms on a 1-connected manifold with finite dimensional cohomology. Assume that $\omega_r \omega_s = 0$. Let*

$$f : \vee S \to X$$

be a smooth map such that $\int_{S^{p_r}} i_r^* f^* \omega_s = \delta_{rs}$.

Then f *induces an injection*

$$\pi_*(\vee S) \otimes R \to \pi_*(X) \otimes R .$$

Proof. We shall use the following fact proved by Sullivan [7], p. 252. Let $g : A \to B$ be a morphism of differential algebra inducing an isomorphism in cohomology. If M is a minimal algebra, then there is a bijection between the homotopy classes (in the algebraic sense) of morphisms of M in A and of M in B .

Consider the following diagram

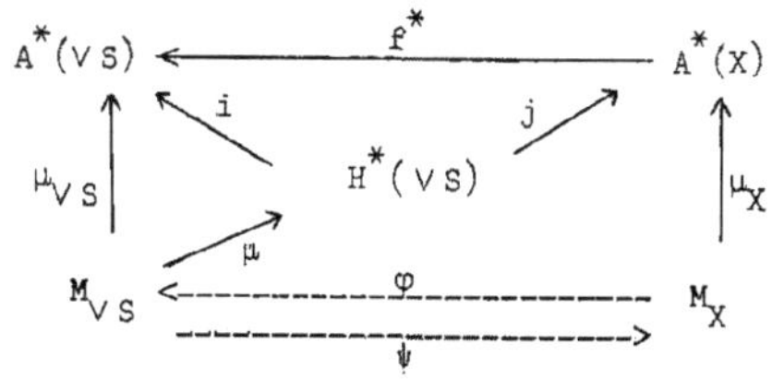

Here $\mu_X : M_X \to A^*(X)$ is a minimal model for the algebra of forms on M, idem for $\mu_{\vee S} : M_{\vee S} \to A^*(\vee S)$. The multiplication in $H^*(\vee S)$ is trivial ; hence there is a morphism of differential algebras $j : H^*(\vee S) \to A^*(X)$ obtained by mapping the generator corresponding to the sphère S^{p_r} on ω_r. Then $f^* \circ j = i$ induces an isomorphism in cohomology. We can first construct the minimal model $\mu : M_{\vee S} \to H^*(\vee S)$ and then define $\mu_{\vee S} = i \circ \mu$.

By what we have just mentioned in the beginning of the proof, there is a map φ such that $\mu_{\vee S} \circ \varphi$ is homotopic to $f^* \circ \mu_X$, and a morphism ψ such that $\mu_X \circ \psi$ is homotopic to $j \circ \mu$. It follows that $\mu_{\vee S} \circ \varphi \circ \psi$ is homotopic to $\mu_{\vee S}$, hence $\varphi \circ \psi$ is homotopic to the identity of $M_{\vee S}$, hence surjective. It follows that ϕ is surjective, hence induces a surjection of $M^+_X / M^+_X \cdot M^+_X$ on $M^+_{\vee S} / M^+_{\vee S} \cdot M^+_{\vee S}$. By duality we get an injection

$$\pi_*(\vee S) \otimes R \to \pi_*(X) \otimes R .$$

Extension to more general spaces X.

We can take for X any 1-connected Kan simplicial set and replace $\vee S$ by a simplicial set, still denoted by $\vee S$, whose geometric realization has the homotopy type of $\vee S$. The theorem is still true in that case, using of course differential forms on simplicial sets. On can first replace X by a subsimplicial set X_o with only one cell in dimension 0 and 1 and express X_o as the union of finite subcomplexes containing the image of f. Then apply the theorem above (with the same proof), and take the direct limit.

5. Applications to some classifying spaces.

The preceding theorem could be applied to classifying spaces like $B\Gamma_n$ or $B\Gamma^C_n$, because the corresponding Gelfand-Fuchs cohomology maps in the forms on those spaces, and many products vanish.

For instance take the case of $F\Gamma^C_n$, the classifying space for trivialized complex foliated microbundles of rank n. In the formulation of the first talk, it is $B\mathfrak{g}$, where $\mathfrak{g}$ is the Lie algebra of formal complex vector fields on C^n [9].

By varying a linear complex vector field on $C^{n+1}-0$, Bott has shown (cf.[8],p.340) that there is a surjective homomorphism φ of $\pi_{2n+1}(F\Gamma_n^C)$ on a product $\prod_\alpha C$ of copies of C indexed by sequences of integers $\alpha = (\alpha_1,\ldots,\alpha_n)$ with $\alpha_1 + 2\alpha_2 + \ldots + n\alpha_n = n+1$, $\alpha_i \geq 0$. The projection on the factor α was obtained by evaluating on S^{2n+1} a form ω_α , coming from the characteristic homomorphism ; the product of two such forms is zero.

Hence, if we choose elements of $\pi_{2n+1}(F\Gamma_n^C)$ whose images under φ form a basis of $\prod_\alpha C$, we get a map f of $\vee_\alpha S^{2n+1}$ in $F\Gamma_n^C$ verifying the hypothesis of the preceding theorem. Hence for each integer $k>0$, we get a lot of non trivial elements of $\pi_{2nk+1}(F\Gamma_n^C)$. By varying f , it is pretty clear by the formulas of § 3 that we get also a surjection of $\pi_{2nk+1}(F\Gamma_n^C)$ on some non trivial vector space over C .

Another example directly related to the result [10] explained by Paul Schweitzer : there is a map $f_o : S^{4k} \to BG\ell_{4k-1}$ on which the Pontryagin class p_k does not vanish. This map lifts to $B\Gamma_{4k-1}$, because $F\Gamma_{4k-1}$ is $(4k-1)$-connected. By the Bott vanishing theorem, p_k^2 vanishes on this lifting f . Applying the theorem of § 2, we see that the Whitehead product

$$[f,f] \in \pi_{8k-1}(F\Gamma_{4k-1})$$

is an element of infinite order.

REFERENCES

[1] N. BOURBAKI — Groupes et Algèbres de Lie, Chapitre 2. Hermann, Paris 1972.

[2] A. HAEFLIGER — Sur la cohomologie de l'algèbre de Lie des champs de vecteurs, à paraître aux Annales de l'Ecole Normale Supérieure.

[3] P.J. HILTON — On the homotopy groups of the union of spheres. J. London Math. Soc. 30 (1955), p. 154-171.

[4] P.J. HILTON and U. STAMMBACH — A course in homological Algebra, Springer Graduate Texts in Math. 4 (1971).

[5] D. QUILLEN — Rational homotopy theory. Annals of Math. 90 (1969), Appendix B, p. 279-295.

[6] N. STEENROD — Cohomology invariants of mappings. Ann. of Math. 50 (1949), p. 954-988.

[7] D. SULLIVAN (with P. DELIGNE, P. GRIFFITHS and J. MORGAN) — Real homotopy theory of Kähler manifolds. Inventiones Math. 29 (1975), 245-274.

[8] BOTT-BAUM — Singularities of holomorphic foliations. J. Differential Geometry 7 (1972), 279-342.

[9] A. HAEFLIGER — Cohomology of Lie Algebras and Foliations, these Proceedings.

[10] P. SCHWEITZER and A. WHITMAN — Pontryagin polynomial residues of isolated foliation singularities, these Proceedings.

On Some Formulas for the Characteristic Classes of Group-actions

Raoul Bott †

1. Introduction. This is an account only of the material of my last two lectures of the Rio conference, as the earlier lectures dealt with well known matters.

My emphasis here then is the study of our foliation invariants in the context of group actions on a manifold, and I will start by showing you the naturality of our basic construction now pays off by extending directly to the equivariant situation.

Recall then the main conclusions of our earlier discussions. Essentially they amounted to this (see also [1], [2] for details):

If we let JM denote the space of jets - based at $0 \in \mathbb{R}^n$ - of diffeomorphisms of $\mathbb{R}^n$ to M, then JM carries an algebra of natural forms which is isomorphic to the cochain algebra $C^*(\mathfrak{a}_n)$ where $\mathfrak{a}_n$ is the Lie algebra of formal vector-fields on $\mathbb{R}^n$.

In short there is a natural arrow:

(1.1) $$C^*(\mathfrak{a}_n) \longrightarrow \operatorname*{Inv}_{\text{Diff } M} \Omega^* JM$$

of $C^*(\mathfrak{a}_n)$ onto the algebra of Diff(M) - invariant forms on JM. Further we saw that although (1.1) induces the 0 - homomorphism in cohomology, it induces an interesting one once we divide JM by the natural action of O_n = the orthogonal group of $\mathbb{R}^n$.

Indeed then (1.1) induces a "basic" homomorphism:

(1.2) $$C^*(\mathfrak{a}_n; O_n) \longrightarrow \operatorname*{Inv}_{\text{Diff } M} \Omega^*(JM/O_n)$$

and this map certainly recaptures all the usual characteristic classes of M.

† The author gratefully acknowledges with thanks partial support from the NSF under Grant MPS 74-11896.

More precisely we saw that:

(1.3) $$H^*(\mathfrak{a}_n, O_n) \simeq H^*(WO_n)$$

where WO_n is the differential algebra given by:

(1.4) $$WO_n = \underline{\mathbb{R}}[c_1, \cdots, c_n] \otimes E(h_1, h_3, \cdots h_k) \quad k \text{ odd and } = n \text{ or } n-1$$

with differential:

$$d\, c_i \otimes 1 = 0 \quad , \quad d(1 \otimes h_i) = c_i$$

and $\underline{\mathbb{R}}$ denoting the quotient of the indicated polynomial ring by the elements of $\dim > 2n$, the c_i having dimension $2i$.

Further we saw - and this is of course quite standard - that

$$H^*(\Omega^*(JM/O_n)) \simeq H^*(M)$$

so that (1.2) induces a natural map

$$H^*(\mathfrak{a}_n, O_n) \longrightarrow H^*(M) \quad ,$$

and finally we identified the image of c_{2i} under this arrow with the Pontryagin classes of M.

The main virtue of this point of view is then, that we see that a manifold determines its own Pontryagin forms naturally on the space

$$\widetilde{M} = JM/O_n \quad ,$$

which I think of as a naturally thickened version of M, and that furthermore these forms are invariant under any diffeomorphism of M.

Now let us put this construction to use when an abstract (i. e., discrete) group Γ, acts on M via diffeomorphism. These data naturally define two semisimplicial manifolds

(1.5) $$M\Gamma : M \leftleftarrows \Gamma \times M \leftleftarrows \Gamma \times \Gamma \times M$$

and

(1.6) $$B\Gamma : * \leftleftarrows \Gamma \leftleftarrows \Gamma \times \Gamma$$

whose geometric realizations then respectively correspond to, the M - bundle, $|M\Gamma|$, induced by the Γ action over the classifying space $|B\Gamma|$ of Γ, and $|B\Gamma|$ itself.

Now if

(1.7) $$X : X_0 \leftleftarrows X_1 \leftleftarrows X_2 \cdots$$

is any semisimplicial manifold the cohomology of its geometric realization $|X|$ can be computed in various ways. First of all the double complex $\Omega^{**}X$

(1.8) $$\Omega^{**}X : \Omega^*(X_0) \xrightarrow{\delta} \Omega^*(X_1) \xrightarrow{\delta} \cdots$$

obtained from (1.7) by applying the de Rham functor Ω^* to each (X_i) and then giving $\Omega^{**}X$ the sum of the de Rham differential and the differential operator δ derived from the simplicial structure, computes $H^*(|X|)$ thus:

(1.9) $$H^*(|X|) = H\{\Omega^{**}(X)\}$$

On the other hand one also has the complex of "compatible forms" on the realization of X:

(1.10) $$|X| = X_0 \cup X_1 \times \Delta_1 \cup X_2 \times \Delta_2 \cup \cdots$$

in the sense of Whitney-Thom-Sullivan. This complex is denoted by $\Omega^*|X|$, and once again one has:

$$(1.11) \qquad H^*(|X|) = H^*(\Omega^*|X|) .$$

For a fine account of all this I refer you to Dupont, [5].
The formula (1.9) is also to be found in [4].

In particular then, we can compute the cohomology of $M\Gamma$ from the double complex

$$(1.12) \qquad \Omega^{**}M\Gamma : \Omega^*(M) \xrightarrow{\delta} \Omega^*(M \times \Gamma) \longrightarrow$$

Furthermore as we will see later, on the first constitutent δ takes the form

$$(1.13) \qquad \delta\omega \mid M \times j = \omega \otimes 1 - j^*\omega \otimes 1 .$$

Hence the natural linear map

$$(1.14) \qquad \Omega^*M \longrightarrow \Omega^{**}M\Gamma$$

<u>becomes a cochain map only on the Γ - invariant forms on M</u>.

In view of this state of affairs it suggests itself to replace $M\Gamma$ by $\widetilde{M\Gamma}$.

Then the arrows (1.2) and (1.14) combine to yield a chain map

$$(1.15) \qquad C^*(a_n, O_n) \longrightarrow \Omega^*(\widetilde{M}) \longrightarrow \Omega^{**}\widetilde{M\Gamma} .$$

On the other hand the homotopy equivalence of M with $\widetilde{M}$ easily implies that of $M\Gamma$ and $\widetilde{M\Gamma}$. Hence in homology (1.15) induces a homomorphism

$$(1.16) \qquad H^*(a_n, O_n) \longrightarrow H^*(|M\Gamma|) .$$

Thus we see that the invariance of our construction immediately yields "equivariant characteristic classes" for a Γ - manifold in $|M\Gamma|$. And here of course, as the cohomology of $M\Gamma$ has no a priori bound, all the classes of WO_n potentially come into play.

Finally if M is a compact Γ - orientable manifold we can define the equivariant characteristic numbers of M by following (1. 16) with integration over the fiber in the fibering

$$\begin{array}{c} M\Gamma \\ \downarrow \pi \\ B\Gamma \end{array} .$$

There results an additive homomorphism

(1. 17) $$H^*(\mathfrak{a}_n; O_n) \longrightarrow H^*(B\Gamma) \quad ,$$

and my aim in the next sections will be to derive some explicit recipes for (1. 17) , and to review some of Thurston's examples in this framework.

2. Formulas for the Godbillon-Vey Class in $H^*(B\Gamma)$. Note that (1. 16) can also be thought of in this manner.

The discreteness of Γ naturally defines a folitation $\mathfrak{F}_\Gamma$ on $M\Gamma$ which is transversal to the fibers in the projection $M\Gamma \to B\Gamma$, and of codimension n . The characteristic classes of $\mathfrak{F}_\Gamma$ integrated over the fiber induce (1. 16) . With this interpretation it suggests itself that the constructions which are known to represent the characteristic classes of folitations should extend to the semisimplicial situation provided only that one has a suitable de Rham theory at hand. Now the double complex $\Omega^{**}X$ does not fit the bill because its multiplication is not anticommutative. On the other hand the compatible complex of Dupont is perfectly suitable and therefore yields explicit recipes quite readily.

Let me start with the Godbillon-Vey class ω, itself. This invariant - corresponding to a generator of $H^3(\mathfrak{a}_1, O_1)$ - is defined on oriented foliations $\mathfrak{F}$ of codimension one, and can be computed according to the algorithm:

(2.1) Let $\mathfrak{F}$ be described as the kernel of a non-degenerate 1 - form φ. Then integrability implies that there exists a 1 - form η with

$$d\varphi = \eta \wedge \varphi .$$

Now set

$$\omega = \eta \wedge d\eta .$$

Then $d\omega = 0$. Further the cohomology class of ω is independent of the choices involved, and represents $\omega(\mathfrak{F})$.

The extension of (2.1) to the semisimplicial case is now immediate. Given a s. s. manifold

$$X : X_0 \leftleftarrows X_1 \leftleftarrows X_2 \cdots$$

a foliation $\mathfrak{F}$ on X is simply a foliation $\mathfrak{F}_k$ on each X_k, such that, all the structure maps:

$$X(\alpha) : X_k \longrightarrow X_{k'}$$

are transversal to $\mathfrak{F}_{k'}$ and induce isomorphisms

$$\mathfrak{F}(\alpha) : \mathfrak{F}_k \longrightarrow X(\alpha)^{-1} \mathfrak{F}_{k'} .$$

Such data then define a foliation $|\mathfrak{F}|$ on the geometric realization $|X|$ of X, in the following manner:

On $X_k \times \Delta^k$, the natural projection

$$X_k \xleftarrow{\pi_L} X_k \times \Delta^k \tag{2.2}$$

induces the foliation $\pi_L^{-1} \circ \mathfrak{F}_k$, and these foliations are compatible under the identifications which assemble the $X_k \times \Delta^k$ to form $|X|$. Precisely we have in mind here the so-called "Fat realization" of $|X|$, which therefore only identifies by the boundary maps. Thus if α is such a boundary map, and

$$\Delta(\alpha) : \Delta^{k-1} \longrightarrow \Delta^k$$

the corresponding map of the $k-1$ simplex Δ^{k-1} onto a face of Δ^k, then the identification corresponding to α is described by the diagram:

$$\begin{array}{ccc} X_k \times \Delta^{k-1} & \xrightarrow{1\times\Delta(\alpha)} & X_k \times \Delta^k \\ {\scriptstyle X(\alpha)\times 1}\downarrow & & \\ X_{k-1} \times \Delta^{k-1} & & \end{array} \tag{2.3}$$

That is, the two images of a point $(p, q) \in X_k \times \Delta^{k-1}$ under the horizontal and vertical arrows are to be identified in $|X|$.

It is clear then that the $\pi_L^{-1} \circ \mathfrak{F}_k$ do define a compatible collection of foliations on $|X|$, and that is precisely what one means by a foliation on $|X|$.

Similarly, one defines the de Rham complex $\Omega^*|X|$ of "compatible forms" on $|X|$ in terms of the diagram (2.3): Thus:

(2.4) A q - form φ on $|X|$ is a collection $\{\varphi_k\}$ of q - forms on $X_k \times \Delta^k$ such that

$$\{1 \times \Delta(\alpha)\}^* \varphi_k = \{X(\alpha) \times 1\}^* \varphi_{k-1} \quad .$$

Finally one has the Dupont extension of the Whitney-Thom-Sullivan Theorem to the effect that

$$H^*(\Omega X) \simeq H(\Omega^{**} X)$$

the isomorphism being induced by "integration over the simplexes"

$$\varphi \mapsto \sum_k \pi_*^L \varphi_k \tag{2.5}$$

where π_L is the projection (2.2) and π_*^L denotes integration over the fiber of π_L. Note that the sum has only q nonzero terms, when φ is of dimension q.

Now then, with all this understood our first remark is:

PROPOSITION 2.6. Let $\mathfrak{F}$ be an oriented codimension 1 foliation on the paracompact simplicial manifold X. Then $\omega(\mathfrak{F}) \in H^*(\Omega|X|)$ can be computed by the Godbillon-Vey algorithm (2.1) provided only we interpret "form" to mean "compatible form" on $|X|$.

The proof of this is quite straightforward, so let me only start the argument and in the process derive an explicit algorithm for $\omega(\mathfrak{F})$ in terms of the structure maps of X.

Consider then an $\mathfrak{F}$ as envisaged in the proposition, and let $\mathfrak{F}_0$ on X_0 be represented as kernel of φ with

$$d\varphi = \eta\varphi \quad .$$

We next try to extend φ and η to $X_1 \times \Delta^1$ in a compatible manner. To fix the notation let 0 and 1 be the vertices of Δ^1 and let x^0 and x^1 be the corresponding barycentric coordinates on Δ^1. Also let

$$\sigma^0 : \Delta^0 \longrightarrow \Delta^1 , \quad \text{and} \quad \sigma^1 : \Delta^0 \longrightarrow \Delta^1$$

be the inclusions sending Δ^0 to 0 and 1 respectively, and let σ_0 and σ_1 be the corresponding maps:

$$X_0 \underset{\sigma_1}{\overset{\sigma_0}{\leftleftarrows}} X_1 \quad .$$

Now by our hypothesis on $\mathfrak{F}$ the forms

$$\sigma_0^* \varphi \quad \text{and} \quad \sigma_1^* \varphi$$

both represent $\mathfrak{F}_1$ in the same orientation. Hence there exists a smooth positive function μ_1 on X_1 such that

(2.6) $$\sigma_1^* \varphi = \mu_1 \, \sigma_0^* \varphi \quad .$$

Using μ_1 we now construct the form

(2.7) $$\varphi_1 = \sigma_0^* \varphi \cdot \mu_1^{x^1} \qquad \text{on} \quad X_1 \times \Delta^1 \quad .$$

Here we have identified the forms on X_1 with their pullback to $X_1 \times \Delta^1$ under π_L and $\mu_1^{x^1}$ is defined by

$$\mu^{x^1}(p, a) = \mu_1(p)^{x^1(a)} \quad .$$

Hence φ_1 restricts to $\sigma_i^*\varphi$ on $\partial(X_1 \times \Delta^1)$ and is compatible. Furthermore the kernel of φ_1 clearly represents $\pi_L^{-1}\mathfrak{F}_1$. Indeed any nonzero multiple of $\sigma_0^*\varphi$ would.

Next we wish to extend η compatibly to $X_1 \times \Delta^1$. For this purpose differentiate (2.7). One obtains

(2.8) $$d\varphi_1 = (\log \mu_1 \cdot dx^1 + x^1 \, d \log \mu_1 + \sigma_0^*\eta) \wedge \varphi_1 \quad .$$

Now the term in the bracket is not compatible with η as it stands. However, it can be modified to become so in the following manner:

By differentiating (2.6) we obtain

$$d\,\sigma_1^*\varphi = \sigma_1^*\eta \cdot \sigma_1^*\varphi = \{d \log \mu_1 + \sigma_0^*\eta\}\,\sigma_1^*\varphi$$

whence

(2.9) $$\sigma_1^*\eta \equiv d \log \mu_1 + \sigma_0^*\eta \mod (\sigma_1^*\varphi) \quad .$$

Hence we may replace $d \log \mu_1$ in (2.8) by $\sigma_1^*\eta - \sigma_0^*\eta$ to obtain:

$$d\varphi_1 = \{\log \mu_1 \, dx^1 + x^1 \, \sigma_1^*\eta_1 + x^0 \, \sigma_0^*\eta\} \wedge \varphi_1$$

and this time the term

(2.10) $$\eta_1 = \{\log \mu_1 \, dx^1 + x^1 \, \sigma_1^*\eta + x^0 \, \sigma_0^*\eta\}$$

is clearly compatible with η.

This construction now extends to all of $|X|$ to yield the following compatible collection φ_k and η_k on $X_k \times \Delta^k$.

For each k, we let $\sigma_0, \sigma_1, \cdots, \sigma_k$ be the maps of $X_k \to X_0$ corresponding to the inclusion of Δ^0 into Δ^k as the $k+1$ vertexes.

Then we define μ_i, $i = 1, \cdots, k$, by

$$\sigma_i^* \varphi = \mu_i \sigma_0^* \varphi$$

and finally set

(2.11)
$$\varphi_k = \sigma_0^* \varphi \left(\prod_{i=0}^{k} \mu_i^{x^i} \right)$$

and correspondingly set

(2.12)
$$\eta_k = \sum_{i=0}^{k} \{\sigma_i^* \eta \cdot x^i + \log \mu_i \cdot dx^i\} \quad .$$

The forms $\eta_k \wedge d\eta_k = \omega_k(\mathfrak{F})$ are therefore again compatible, closed, and give an algorithm for computing $\omega(\mathfrak{F})$ in $H^*(|X|)$.

At this stage one may of course return to the more economical complex $\Omega^{**}(X)$ by integrating over the simplices. Then

(2.13)
$$\pi_*^L \omega(\mathfrak{F}) = \omega^{3,0} + \omega^{2,1} + \omega^{1,2} + \omega^{0,3}$$

with $\omega^{i,j} \in \Omega^i(X_j)$ and using (2.13) one obtains explicit formulae. For example:

(2.14)
$$\omega^{3,0} = \eta \wedge \varphi ,$$

while $\omega^{2,1}$ is obtained by integrating $\eta_1 \wedge d\eta_1$ over the 1 - simplex Δ^1. Thus this component is given by:

$$\int_{\Delta^1} (x^0 \sigma_0^*\eta + x^1 \sigma_1^*\eta + \log \mu_1 \, dx^1) \wedge (dx^1(\sigma_1^*\eta - \sigma_0^*\eta) + d \log \mu_1 \, dx^1$$

$$+ x^0 \, d\sigma_0^* + x^1 \, d\sigma_1^*\eta) \quad .$$

Only the terms involving a dx survive and hence with a little algebra one arrives at

$$\omega^{2,1} = \sigma_1^*\eta \wedge \sigma_0^*\eta - d\{\log \mu_1(\frac{\sigma_1^*\eta + \sigma_0^*\eta}{2})\} \tag{2.15}$$

To obtain the next term we proceed similarly with $\eta_2 \wedge d\eta_2$. The result is

(2.16)

$$\omega^{1,2} = \sum_{i,j=0}^{2} \int_{\Delta^2} \log \mu_i \, (d \log \mu_j - \sigma_j^*\eta) \, dx^i \, dx^j \qquad 0 \leq i, j \leq 2$$

$$= \sum_{0 \leq i < j \leq 2} (-1)^{i-j+1} \log u_i (d \log \mu_j - \sigma_j^*\eta)$$

Finally $\omega^{0,3}$ vanishes because there is no term involving three dx's in $\eta_3 \wedge d\eta_3$.

So far I have discussed only the original Godbillon-Vey class ω for foliations of codimension 1. If $\mathfrak{F}$ is orientable but has higher codimension say q - their class ω is generalized to $\mathfrak{F}$, by again describing $\mathfrak{F}$ as the kernel of a decomposable form φ, this time of dimension q. Then integrability again implies that

$$d\varphi = \eta \wedge \varphi$$

but now it is the form

$$\omega(\mathfrak{F}) = \eta \wedge (d\eta)^q$$

which is an invariant of $\mathfrak{F}$ and thus gives the extended Godbillon-Vey class in $H^{2q+1}(M)$. In WO_q this class then represents $h_1 c_1^q$.

In the semisimplicial situation this extension works equally well and (2. 12) again yields an explicit algorithm for the computations of $\omega(\mathfrak{F})$. Note by the way that in codimension q, only components of the type;

(2. 17) $$\pi_*^L \omega(\mathfrak{F}) = \omega^{2q+1,0} + \cdots + \omega^{q,q+1}$$

survive because $\eta_k \wedge d\eta_k{}^q$ can contain no more than $(q+1)$ dx's for any k. Furthermore the extremal terms are easy to compute from (2. 12):

(2. 18)
$$\omega^{2q+1,0} = \eta \wedge (d\eta)^q$$
$$\omega^{q,q+1} = \int_{\Delta^q} (\log \mu_i \, dx^i) \{(d \log \mu_j - \sigma_j^* \eta) \, dx^j)^q\} \quad .$$

There are corresponding formulae for the other classes in $H^*(WO_n)$ but they are harder to write down explicitly, because, as in the case of just a single foliation $\mathfrak{F}$ on M, one has to bring in a comparison of torsion-free and Riemannian connections.

However the compatible complex again enables one to extend the single-space recipes to the semisimplicial case in a more or less straightforward manner, but rather than discussing these let me now specialize our formulae to the case of a group action. We therefore have to recall briefly how $M\Gamma$ and $B\Gamma$ are constructed and why $M\Gamma$ carries a natural foliation.

Recall first, that if C is a a category, then it defines a natural semisimplicial object

(2. 19) $$BC : O(C) \leftleftarrows \mathfrak{m}_1(C) \equiv \mathfrak{m}_2(C)$$

starting with the objects $O(C)$ of C, and $\mathfrak{m}_i(C)$ being the i-tuples of composable morphisms in C. Thus a typical element in $\mathfrak{m}_k(C)$ is given by the diagram:

(2.20)
$$\cdot \xleftarrow{\gamma_1} \cdot \xleftarrow{\gamma_2} \cdot \;\cdots\; \cdot \xleftarrow{\gamma_k} \cdot$$

the γ ' s being arrows in C. The boundary maps in BC are then given by sending such a diagram into one with either an end arrow deleted, or two consecutive one composed. For example

(2.21)
$$\partial_0(\gamma_1, \gamma_2) = \gamma_2$$
$$\partial_1(\gamma_1, \gamma_2) = \gamma_1 \circ \gamma_2$$
$$\partial_2(\gamma_1, \gamma_2) = \gamma_1$$

This construction applied to the category determined by the group Γ yields $B\Gamma$. When applied to the category Γ_M determined by the action of Γ on M, it yields $M\Gamma$: Thus

$$M\Gamma \equiv B\Gamma_M$$

where the category Γ_M has as objects the manifold M, and as morphisms pairs (γ, m) with $\gamma \in \Gamma$, $m \in M$. Finally the composition

(2.22)
$$(\gamma_1, m_1) \circ (\gamma_2, m_2) = (\gamma_1 \circ \gamma_2, m_2)$$

exists only if $\gamma_2(m_2) = m_1$ and is then given by the RHS of (2.22).

Pictorially these are then best described by arrows of the form

(2.23)
$$\gamma(m) \xleftarrow{\gamma} m$$

In any case, all of this granted it is easy to see that the following proposition holds.

PROPOSITION 2.24. Let $\mathfrak{F}$ be a foliation on M, such that Γ acts transversally to $\mathfrak{F}$ on M. Then $\mathfrak{F}$ extends to a foliation $\mathfrak{F}\Gamma$ on $M\Gamma$.

In particular the foliation of M by points induces a natural foliation $\mathfrak{F}_M$ on $M\Gamma$.

Note of course that $\mathfrak{F}_M$ then corresponds to the foliation by points of $M\Gamma$ considered as a simplicial manifold in the obvious sense.

When we specialize our formulae to $\mathfrak{F}_M$, matters simplify to some extent as now η can clearly be taken to be zero. If we furthermore follow our recipe, in the case when M is oriented and compact, and integrate $\omega(\mathfrak{F}_M)$ over the fiber in the projection

$$M\Gamma \xrightarrow{\pi} B\Gamma$$

then - M being q - dimensional - only $\omega^{q,q+1}$ enters into the computation. In this manner we obtain the following algorithm for

$$\pi_* \, \omega(\mathfrak{F}_M) \quad .$$

PROPOSITION 2.25. The characteristic "number" $\pi_* \, \omega(\mathfrak{F}_M) \in H^{q+1}(B\Gamma)$ is represented by the cocycle:

$$(2.26) \quad \omega(\gamma_1, \cdots, \gamma_{q+1}) = \int_{\Delta^{q+1} \times M} (\log \mu_i \, dx^i)(d \log \mu_j \, dx^j)^q$$

where the μ_i are functions on M, determined in the following manner:

Choose a volume φ on M, and then define μ_i by:

$$(2.27) \qquad (\gamma_i \gamma_{i+1} \cdots \gamma_{q+1})^* \varphi = \mu_i \varphi \quad .$$

Remark. (1) Thurston has been aware of this formula all along although his conventions and his setting might have differed somewhat. For those of us less geometrically inspired it was however not so obvious. On the other hand - and this is the main point of my remarks - in the context of the compatible complex the formula does become easily accessible.

(2) The following geometrical interpretation of 2.25 was pointed out to me by Thurston quite recently. If $y_1, \cdots, y_{q+1}$ are coordinates in $\mathbb{R}^{q+1}$ then the form

(2.28) $$\Psi = \Sigma(-1)^i y_i \, dy_1 \wedge \cdots \widehat{dy_i} \wedge \cdots dy_n$$

clearly satisfies the identity

$$d\Psi = n \text{ volume in } \mathbb{R}^n .$$

On the other hand, up to a universal constant (2.26) clearly gives the integral of Ψ pulled back to M under the map

$$f_\Gamma : M \longrightarrow \mathbb{R}^{q+1}$$

defined by

(2.29) $$y^i \circ f_\Gamma = \log(\mu_i)$$

Thus our formula for the Eilenberg MacLane cocycle $\omega(\mathfrak{F})$ can be written;

(2.30) $$\omega(\mathfrak{F})(\gamma_1, \cdots, \gamma_{q+1}) = \text{volume enclosed by } f_\Gamma M \text{ in } \mathbb{R}^{q+1} .$$

3. The Infinitesimal Case. When the action of Γ on a foliation $\mathfrak{F}$ over M comes by restriction of an action of a connected Lie group G containing Γ, the characteristic classes of the Γ - action can be described in Lie-algebra terms. Such a description is often more easy to handle, and our aim in this section is therefore to describe an infinitesimal analogue of (2. 18) in this framework. The technique involves the lifting of $\mathfrak{F}$ to to $(G/K) \times M$, with K a maximal compact subgroup of G, and is thus typical of the Van-Est theorems and theorems on continuous cohomology, as well as of the whole approach of Kamber and Tondeur to foliated bundles. See [7].

However I will not pursue these aspects, but rather head as quickly as possible towards the infinitesimal formula for $\omega(\mathfrak{F})$.

Recall then, first of all, that G/K is contractible:

(3. 1) $$G/K \simeq pt \quad .$$

Hence if we consider the natural Γ action on G/K then (3. 1) extends to a homology equivalence of the quotient of G/K by the action of Γ (denoted $[G/K]\Gamma$) with $B\Gamma$,"

(3. 2) $$[G/K]\Gamma \simeq B\Gamma$$

Thus $[G/K]\Gamma$ is a thickened version of $B\Gamma$. Now consider the edge homomorphism in $[G/K]\Gamma$,

(3. 3) $$\operatorname*{inv}_{\Gamma} \Omega^*(G/K) \to \Omega^{**}([G/K]\Gamma) \quad .$$

Because $\Gamma \subset G$, the left hand side certainly contains the subcomplex of G - invariant forms on G/K, which is finally identified with the relative Lie algebra complex $\Omega^*(\underline{g}; K)$. Thus we have

$$\Omega^*(\underline{g};K) \equiv \operatorname*{Inv}_G \Omega^*(G/K) \subset \operatorname*{Inv}_\Gamma \Omega^*(G/K) \tag{3.4}$$

combining (3.4) with (3.3) and (3.2) gives rise to an arrow †

$$\iota_* : H^*(\underline{g};K) \to H^*(B\Gamma) \quad , \tag{3.5}$$

and our first aim is then to show that:

PROPOSITION 3.6. <u>In the situation envisaged the characteristic classes of the</u> Γ <u>action lift naturally to</u> $H^*(\underline{g};K)$.

To see this result one first has to understand the following diagram of G spaces:

$$G \underset{K}{\times} M \xrightarrow[\approx]{t} (G/K) \times M \xrightarrow{\pi} M \quad , \tag{3.7}$$

where π is the natural product projection, and the action of G on $(G/K) \times M$ is the product action. On the other hand G <u>acts on</u> $G \underset{K}{\times} M$ <u>purely on the left, and</u> $G \underset{K}{\times} M$ <u>is</u> defined as the quotient

$$G \underset{K}{\times} M = (G \times M)/K \quad , \tag{3.8}$$

with K acting by

$$(g,m) \cdot k = (gk, k^{-1}m) \quad . \tag{3.9}$$

Finally the twist map t is induced by

$$\tilde{t} : G \times M \to G \times M \tag{3.10}$$

sending

$$(g,m) \quad \text{to} \quad (g, g\cdot m) \quad . \tag{3.11}$$

---

† In [5] and also independently in a recent paper of Shulman and Tischler, this arrow is explicitly described on the chain level.

It is then clear that $\tilde{t}$ is a diffeomorphism which sends the K action (3.9) into the K action $(g, m) \to (gk, m)$ and so induces the equivalence t of (3.7). The virtue of this twisting map t, is of course that the G - action on $G \times_K M$ is given by the left translation of G. It follows that the G - invariant forms on $G \times_K M$ can be identified with the forms on $G \times M$, which are

(1) invariant under left translation of G by G, and

(2) K - basic under the action (3.8) of K on $G \times M$. Thus:

$$\text{(3.12)} \qquad \operatorname{Inv}_G \Omega^*(G \times_K M) \simeq K \text{ basic forms in } \Omega^*(\underline{g}) \otimes \Omega^* M \quad .$$

In any case, the plan of procedure is now suggested by the diagram:

$$\text{(3.13)} \qquad \begin{array}{ccccc} \operatorname{Inv}_G \Omega^*(G \times_K M) & \to & \Omega^{**}(G \times_K M\Gamma) & \approx & \Omega^{**}(M\Gamma) \\ \downarrow \pi_* & & & & \downarrow \pi_* \\ \operatorname{Inv}_G \Omega^*(G/K) & \to & \Omega^*(G/K\Gamma) & \approx & \Omega^{**}(B\Gamma) \end{array}$$

where the lower line induces (3.5), and π_* denotes integration over the fiber. The homotopy equivalence in the upper line is of course again a consequence of the contractibility of G/K. In view of (3.13) our lifting problem clearly amounts to realizing the characteristic classes of $\mathfrak{F}$ lifted to $G \times M$, by G - invariant forms. Thus we need the following.

PROPOSITION 3.14. The characteristic classes of $\mathfrak{F}$ admit a natural lifting to the complex (3.12).

Let me carry out the proof, but again only for our Godbillon-Vey class $\omega(\mathfrak{F})$. Consider then the pull-back $\tilde{\mathfrak{F}}$ of $\mathfrak{F}$ to $G \times M$, under the map

$$\text{(3.15)} \qquad G \times M \xrightarrow{\pi \circ \tilde{t}} M \quad .$$

To describe $\tilde{\mathcal{F}}$ let us identify the tangent space of $G \times M$, at (g, m) with $\underline{g} \oplus T_m M$, using the left invariant vector-fields on G to identify G_g with $\underline{g}$, and using the projections for the direct sum decomposition. Also if $x \in \underline{g}$ is an element of the Lie algebra of G, let $\dot{x} \in \Gamma(TM)$ be the corresponding infinitesimal motion on M, induced by the action of G on M.

Precisely if e^{tx} is the one-parameter subgroup generated by $x \in \underline{g}$, then

$$\dot{x}_m = \text{tangent of } e^{tx}m \text{ at } m \quad . \tag{3.16}$$

With this understood, the kernel of $\tilde{\mathcal{F}}$ is described by:

$x + y \in T_{(g,m)}(G \times M)$ <u>is in</u> $\tilde{\mathcal{F}}$ <u>if and only if</u>

$$\dot{x}_m + y_m \in \mathcal{F} \quad . \tag{3.17}$$

Indeed the curve (ge^{tx}, m) goes over into $ge^{tx}m$ under our map, and hence its tangent at $t = 0$ goes to $g_*\dot{x}_m$. On the other hand y goes to $g_* y_m$. Hence $g_*(\dot{x}_m + y_m) \in \mathcal{F}$. But G preserves the foliation so that (3.17) follows.

An immediate corollary of (3.17) is the following:

PROPOSITION 3.18. <u>Let the foliation</u> $\mathcal{F}$ <u>be described as the kernel of the decomposable</u> q <u>form</u> φ. <u>Then</u> $\tilde{\mathcal{F}}$ <u>admits a natural representation as the kernel of a decomposable form</u> $\tilde{\varphi} \in \Omega^*(\underline{g}) \otimes \Omega^*(M)$. <u>Furthermore, in the natural double grading of this complex</u> $\tilde{\varphi}$ <u>has components:</u>

$$\tilde{\varphi} = \tilde{\varphi}^{q,0} + \cdots + \tilde{\varphi}^{0,q}$$

with:

$$\tilde{\varphi}^{0,q} = \varphi \quad .$$

(3. 19) $$\tilde{\varphi}^{1,q-1} = \sum_{\alpha} x'_{\alpha} \wedge \iota(\dot{x}_{\alpha})\,\varphi$$

where x_{α} runs over a basis of g, and x'_{α} over a dual base in g^*.

Proof. This is purely a linear algebra matter. If θ is a 1 - form with $\mathfrak{F}$ in its kernel, then

(3. 20) $$\tilde{\theta} = \theta + \sum_{\alpha} x'_{\alpha}\, \iota(\dot{x}_{\alpha})\,\theta$$

will have $\tilde{\mathfrak{F}}$ in its kernel. Indeed for any $x + y$ subject to (3. 17) we then have $\iota(x+y)\tilde{\theta} = \theta(y) + \theta(\dot{x}) = 0$.

Hence if $\theta^1 \wedge \cdots \wedge \theta^q = \varphi$ locally, then $\tilde{\theta}^1 \wedge \cdots \wedge \tilde{\theta}^q$ describes $\tilde{\mathfrak{F}}$ locally, and expanding this product clearly yields (3. 19).

To proceed further we need to compute $d\tilde{\varphi}$ and express it as $\tilde{\eta} \wedge \tilde{\varphi}$ in the double complex $\Omega^*(g) \otimes \Omega^*(M)$.

For this purpose, let us set

$$\iota(\dot{x}_{\alpha})\varphi = \varphi_{\alpha}$$

and use the double index summation convention, so that

(3. 21) $$\tilde{\varphi} = \varphi + x'_{\alpha} \wedge \varphi_{\alpha} + \cdots$$

describes the "beginning" of $\tilde{\varphi}$.

Also, let us assume that on M the integrability of $\mathfrak{F}$ is expressed by

(3. 22) $$d\varphi = \eta \wedge \varphi$$

where η is a global 1 - form. Then I claim that:

PROPOSITION 3.23. The η of (3.22) lifts naturally to one $\tilde{\eta}$ in $\Omega^*\underline{g} \otimes \Omega^* M$ such that

$$d\tilde{\varphi} = \tilde{\eta} \wedge \tilde{\varphi} \quad .$$

Further the $\tilde{\eta}$ is given by

(3.24) $$\tilde{\eta} = \eta - x'_\alpha \{\mu(x_\alpha) - \eta(\dot{x}_\alpha)\} \text{ where } \mu(x) \text{ is defined by } (3.27).$$

Proof. Differentiating (3.21) yields

(3.25) $$d\tilde{\varphi} = d\varphi - x'_\alpha \wedge d\varphi_\alpha + \cdots \quad .$$

Further

(3.26) $$d\varphi_\alpha = d\iota(\dot{x}_\alpha)\varphi = \mathcal{L}(\dot{x}_\alpha)\varphi - \iota(\dot{x}_\alpha)d\varphi.$$

where $\mathcal{L}(\dot{x}_\alpha)$ is the Lie derivative in the direction $\dot{x}$. Because G preserves $\mathcal{F}$, $\mathcal{L}(\dot{x})$, $x \in \underline{g}$ must preserve φ up to multiples, whence

(3.27) $$\mathcal{L}(\dot{x})\varphi = \mu(x)\varphi \ , \ x \in \underline{g}\ , \ \text{ with } \ \mu(x) \in \Omega^0(M) \quad .$$

This $\mu(x)$ is, of course, the infinitesimal analogue of the $\mu(\sigma)$ in Section 2.

In any case, combining (3.25), (3.26), (3.27) with (3.22) one obtains the formula

$$d\tilde{\varphi} = \eta \wedge \varphi - \mu(x_\alpha)x'_\alpha \wedge \varphi$$
$$+ x'_\alpha \wedge \{\eta(\dot{x}_\alpha)\varphi - \eta \wedge \varphi_\alpha\} + \cdots$$

which, up to terms of order ≥ 2 in the $\underline{g}$ direction, is given by

(3.28) $$d\tilde{\varphi} = \{\eta - x'_\alpha(\mu(x_\alpha) - \eta(\dot{x}_\alpha)\} \wedge \tilde{\varphi} \quad .$$

But as $\widetilde{\eta}$ exists and clearly is the sum of forms of type $(1,0)$ and $(0,1)$ this equation fixes $\widetilde{\eta}$. Q.E.D.

To assemble the pieces, we shall have to determine whether $\widetilde{\varphi}$ and $\widetilde{\eta}$ are K basic in our complex.

In general this will, of course, not be the case. However, by averaging over K, we can arrange it that both the φ and the η of our discussion are invariant under the action of K, i.e., that infinitesimally

(3.29) $$\mathcal{L}(\dot{x})\varphi = 0 \quad ; \quad \mathcal{L}(\dot{x})\eta = 0 \quad \text{for} \quad x \in \underline{k}$$

and under this hypothesis we have the following.

PROPOSITION 3.30. The condition (3.29) implies that $\widetilde{\varphi}$ and $\widetilde{\eta}$ are K basic[†] relative to the action of K on $G \times M$.

The proof of this fact is a straightforward check (though not quite trivial) which I will take up in greater generality at another time.

At this stage, we are ready to give an infinitesimal recipe for $\omega(\mathfrak{F})$.

Indeed, expanding $\widetilde{\eta}(d\widetilde{\eta})^q$, will have all possible type of components:

$$\omega(\mathfrak{F}) = \omega^{2q+1,0} + \cdots + \omega^{0,2q+1} ,$$

of which the simplest are given by:

(3.31) $$\omega^{0,2q+1} = \eta \cdot d\eta^q ,$$

(3.32) $$\omega^{q+1,q} = - x'_{\alpha_1} \wedge \cdots x'_{\alpha_{q+1}} \nu_{\alpha_1} d\nu_{\alpha_2} \wedge \cdots d\nu_{\alpha_{q+1}}$$

where we have now set

[†] The K basic forms are those annihilated by the vector fields along the orbit of the K action and invariant under that action.

(3.33) $$\nu(x) = \mu(x) - \eta(\dot{x}) \qquad x \in \underline{g}$$

and have abbreviated $\nu(x_\alpha)$ to ν_α.

If we think of $\Omega^*(\underline{g}) \otimes \Omega^*(M)$ as the complex $\Omega^*(\underline{g}\,;\Omega^*(M))$ of forms on $\underline{g}$ with values in Ω^*M, then (3.33) can be thought of as a 1 - form on $\underline{g}$ with values in $\Omega^o(M)$:

$$\nu \in \Omega^1(\underline{g}\,;\Omega^o(M)) \quad ,$$

and there (3.32) takes the form

(3.34) $$\omega^{q+1,q}(x_1, \cdots, x_{q+1}) = \frac{1}{(q+1)} \Sigma\, (-1)^i \nu(x_i)\, d\nu(x_1) \cdots d\nu(\hat{x}_i) \cdots d\nu(x_{q+1}) \quad .$$

Hence we get the following corollary, which is an infinitesimal analogue of 3.25.

PROPOSITION 3.35. Suppose $Z \subset M$ is a cycle of dimension q on M. Then

(3.36) $$\int_Z \omega(\mathfrak{F}) \in H^{q+1}(\underline{g}\,;K)$$

is represented by the cocycle

(3.37) $$\omega_Z(\mathfrak{F})\,(x_1, \cdots, x_q) = \int_Z \nu(x_1)d\nu(x_2) \cdots d\nu(x_{q+1})$$

where $\nu(x)$ is defined by

(3.38) $$\{\mathcal{L}(x) - \eta(\dot{x})\}\omega = \nu(x)\omega \quad .$$

4. On the Examples of Thurston and Heitsch. The history of examples of foliations with varying classes $\omega(\mathfrak{F})$ is roughly as follows.

In the complex analytic case, I had observed already before 1970 that the foliation:

$$(4.1) \qquad \mathfrak{F}_\lambda = \{\lambda_1 z_1 \frac{\partial}{\partial z_1} + \lambda_2 z_2 \frac{\partial}{\partial z_2}\}$$

on $\mathbb{C}^2 - 0$, had for its g. v. invariant:

$$\int_{S^3} \omega(\mathscr{F}_\lambda) = c\,\{\frac{\lambda_1}{\lambda_2} + \frac{\lambda_2}{\lambda_1} - 2\} ,$$

and thus varied continuously with λ.

At that time, I thought that the corresponding real invariant would always vanish. However, soon thereafter in 1971, the paper of Godbillon-Vey appeared with the Roussarie example of the foliation $\mathfrak{F}$ on $\Gamma\backslash SL(2;\mathbb{R})$ induced by the Lie algebra of triangular matrices in $SL(2;\mathbb{R})$, Γ being a discrete subgroup with compact quotient space.

Thereafter, Thurston produced his examples. In particular he constructed examples of a family of actions of a group Γ acting on S^1, such that the corresponding g. v. number $\in H^2(B\Gamma)$ varied continuously. In an appendix, Robert Brooks has written up the details treating this example with the formula (2.25) much like Dupont in [5] treated the Euler class of flat bundles. Here let me just outline and comment on the plan of this very ingenious example.

We start by observing that $SL(2;\mathbb{R})$ acts on S^1 in the classical manner

$$(4.1) \qquad z \longrightarrow \frac{az+b}{cz+d} , \quad |z| = 1, \begin{pmatrix} a & b \\ c & d \end{pmatrix} \in SL(2,\mathbb{R}) .$$

Hence for any $\Gamma \subset SL(2,\mathbb{R})$ there is a natural action of Γ on S^1.
Where $\Gamma\backslash SL(2,\mathbb{R})$ is compact, $H^2(\Gamma;\mathbb{R}) \neq 0$ and the g. v. class of the action will also be nonzero. On the other hand, as we let Γ vary in a continuous family of such subgroups,

the corresponding characteristic class does not vary. Thus, the moduli of Riemann surfaces do not furnish varying examples.

Thurston therefore twisted these homogeneous actions on S^1 in the following manner.

Consider the double cover

$$S^1 \xrightarrow{\pi} S^1$$

given by sending z to z^2; $|z| = 1$. Then every diffeomorphism f of S^1, admits precisely two liftings $\tilde{f}$ relative to π. Thus,

$$\tilde{f} : S^1 \longrightarrow S^1$$

is a diffeomorphism of S^1, with

$$\pi \circ \tilde{f} = f \circ \pi \ .$$

On the double cover $\mathrm{Diff}^{(2)} S^1$ of $\mathrm{Diff}\, S^1$ the function $f \longrightarrow \tilde{f}$ now becomes single valued and defines a homomorphism:

(4.2) $$\mathrm{Diff}^{(2)} S^1 \xrightarrow{\Psi_2} \mathrm{Diff}\,(S^1)$$

Note that if f is lifted to an "equivariant map": $\underline{f}(x + 2\pi) = \underline{f}(x) + 2\pi$,

$$\underline{f} : \mathbb{R} \longrightarrow \mathbb{R}$$

then $\tilde{f}$ is represented by:

$$\underline{\tilde{f}}(x) = 1/2\, \underline{f}\,(2x) \quad \text{or} \quad 1/2\, \{\underline{f}(2x) + \pi\} \ .$$

Hence, in particular, if f is a rotation by α and hence represented by

$$x \longrightarrow x + \alpha$$

then $\tilde{f}$ is represented by:

(4.3) $$x \longrightarrow x + \alpha/2, \quad \text{and} \quad x \longrightarrow x + \frac{\alpha}{2} + \pi \quad .$$

It follows that if the map

$$SL(2,\mathbb{R}) \longrightarrow PSL(2,\mathbb{R}) \longrightarrow \text{Diff } S^1$$

given by (4.1), is lifted to a map of $SL(2,\mathbb{R})$ to $\text{Diff}^{(2)}(S^1)$ and then followed by Ψ_2, there results a homomorphism of

(4.4) $$SL(2,\mathbb{R}) \underset{SO(2)}{*} SL(2,\mathbb{R}) \xrightarrow{1 \times \Psi_2} \text{Diff }(S^1)$$

where on the left we have in mind the free product amalgamated along the rotations according to (4.3), and it is this action which gives rise to Thurston's example. More precisely, he takes the Γ represented by generators

$$\{X, Y, Z, W\}$$

and with the relation

(4.5) $$[X,Y] = [Z,W] \quad ,$$

chooses $X, Y \in SL(2;\mathbb{R})$ so that $[X,Y]$ is a rotation by $0 < \alpha < \pi$, (see Appendix for details), and also $Z', W' \in SL(2,\mathbb{R})$ so that $[Z',W']$ is a rotation by 2α. Then lifting Z' and W' and applying Ψ_2, one obtains Z, W in $\text{Diff } S^1$, which clearly satisfy the relation (4.5).

Now varying α, Thurston obtains his example. Note that as this example is obtained by amalgamation of two infinitesimal situations, it cannot be directly treated by our infinitesimal method, although, as Bob Brooks point out our global and infinitesimal cocycles agree where they should.

More recently Thurston has found examples of varying the higher g.v. classes by actions of certain Γ's on the spheres. Thus, his examples actually vary the characteristic numbers of certain Γ actions.

During the Rio conference, James Heitsch suggested a different approach to these examples, which quite recently has enabled him to vary a large number of characteristic classes independently (see [6]).

In our terminology, Heitsch passes from the sphere, where Thurston worked, to a foliation $\mathfrak{F}_\lambda$ on $\mathbb{R}^n - 0$ of the type I used in the complex case, but here he starts out with a sufficiently special λ so that $\mathfrak{F}_\lambda$ admits a large group of automorphism. Let me conclude by taking up the first instance of his construction to vary $h_1c_1^3 \in H^4(B\Gamma)$.

We will use our infinitesimal recipe for this purpose, so recall first of all that the homomorphism

(4.6) $$H^*(\underline{g}; K) \longrightarrow H^*(B\Gamma)$$

is injective for any semi-simple Lie group and any discrete subgroup Γ with $\Gamma\backslash G$ compact. Indeed Γ will then have a subgroup of finite index $\Gamma' \subset \Gamma$ such that Γ' acts freely on G/K. The natural map

(4.7) $$H^*(\underline{g}; K) \longrightarrow H^*(\Gamma'\backslash G/K)$$

is then injective in the top dimension and both sides satisfy Poincaré duality. Hence (4.7) is injective, but $\Gamma'\backslash G/K \simeq B\Gamma'$, and $B\Gamma'$ and $B\Gamma$ have equal rational cohomology. Q. E. D.

Finally recall that by a theorem of Borel's any semi-simple G admits a Γ with $\Gamma\backslash G$ compact. Thus for our purposes, varying the infinitesimal class with a semi-simple group G also varies the class in some $B\Gamma$.

With these remarks we are ready to take up the Heitsch example.

Let then $\mathfrak{F}_\lambda$ be generated by X_λ in $\mathbb{R}^4 - 0$ where

$$X_\lambda = \sum_{i=1}^{2} \lambda_i \left(x_i \frac{\partial}{\partial x_i} + y_i \frac{\partial}{\partial y_i} \right). \tag{4.8}$$

Then the natural action of $SL(2;\mathbb{R})$ on the (x_i, y_i) space defines an action of

$$G = SL(2,\mathbb{R}) \times SL(2,\mathbb{R})$$

on $\mathbb{R}^4$, which obviously preserves X_λ and hence $\mathfrak{F}_\lambda$.

The infinitesimal action of $\underline{g}$ on $\mathbb{R}^4$ is therefore generated by

$$u_i = x_i \frac{\partial}{\partial x_i} - y_i \frac{\partial}{\partial y_i} \quad ; \quad v_i = x_i \frac{\partial}{\partial y_i} + y_i \frac{\partial}{\partial x_i} \tag{4.9}$$

and

$$h_i = x_i \frac{\partial}{\partial y_i} - y_i \frac{\partial}{\partial x_i} \quad .$$

Thus the h_i generate the action of K, and a class in $H^4(\underline{g};K)$ is determined by its value on $u_1 \wedge v_1 \wedge u_2 \wedge v_2$.

Let us now apply the infinitesimal recipe to $\mathfrak{F}_\lambda$. For φ we may choose

$$\varphi = \iota(X_\lambda)v \quad , \quad v = dx_1\, dy_1\, dx_2\, dy_2 \quad . \tag{4.10}$$

Then

$$d\varphi = \mathcal{L}(X_\lambda)v = 2(\lambda_1 + \lambda_2)v \quad .$$

Hence an admissible η is given by

$$\eta = (\lambda_1 + \lambda_2) \cdot \frac{\Sigma \lambda_i^2 \, dr_i^2}{\Sigma \lambda_i^2 \, r_i^2} \quad , \quad i = 1,2 \tag{4.11}$$

where we have set

$$r_i^2 = x_i^2 + y_i^2 \quad . \tag{4.12}$$

Indeed η is clearly invariant under K, and satisfies the relation

$$d\varphi = \eta \wedge \varphi \quad .$$

Further note that φ is invariant under all of G, so that the ν of (3.33) is entirely given by the formula

$$\nu(x) = \eta(x) \quad .$$

Now by direct computation

$$\begin{aligned} u_i \cdot r_i^2 &= 2(x_i^2 - y_i^2) \\ v_i \cdot r_i^2 &= 4 \, x_i \, y_i \end{aligned} \tag{4.13}$$

hence

(4.14)

$$\nu(u_i) = -\eta(u_i) = \frac{\left(\lambda_1+\lambda_2\right) 2\lambda_i \left(x_i^2 - y_i^2\right)}{\Sigma \lambda_i^2 r_i^2}$$

$$\nu(v_i) = -\eta(v_i) = \frac{\left(\lambda_1+\lambda_2\right) 4\lambda_i x_i y_i}{\Sigma \lambda_i^2 r_i^2}$$

Now (3.37) together with (2.28) yield the formula:

$\omega(\mathfrak{F}_\lambda)(u_1,v_1,u_2,v_2)$ = <u>volume enclosed by the map</u> $S^3 \longrightarrow R^4$ <u>given by the four functions</u>

$\nu(u_1)$, $\nu(v_1)$, $\nu(u_2)$, $\nu(v_2)$

<u>on</u> $S^3 \subset R^4$.

Clearly this map is homogeneous of degree zero, hence we may change coordinates from x_i to x_i/λ_i, and setting

(4.15) $$z_i = x_i + \sqrt{-1}\, y_i$$

we see that the volume enclosed by the map in question will be proportional to $(\lambda_1+\lambda_2)^4/(\lambda_1\lambda_2)^2$ times the volume enclosed by the unit sphere under the map,

(4.16) $$\{z_1, z_2\} \longrightarrow \{z_1^2, z_2^2\} / |z_1|^2 + |z_2|^2 ,$$

which is easily seen to be nonzero. <u>Hence</u>

$$\int \omega(\mathfrak{F}_\lambda)(u_1,v_1,u_2,v_2) = \text{const.} \frac{\left(\lambda_1+\lambda_2\right)^4}{\left(\lambda_1\lambda_2\right)^2}$$

<u>and therefore varies with</u> λ.

In higher even dimensions this method of Heitsch's works equally well and leads to an independent variation of all the $h_1 c^{\alpha}$ classes. In odd dimensions the argument is a little more subtle. However, he can also treat this case by construction which - on the sphere - goes back to Thurston.

All in all then Thurston and Heitsch have set us well on the way of showing that all the potential classes of $H^*(\mathfrak{a}_n; \mathfrak{o}_n)$ are non-trivial, independent and, in the appropriate cases, variable. For classes involving many h's, corresponding independence theorems were first obtained by Kamber and Tondeur [7].

Appendix

by

Robert Brooks

In this appendix, we will evaluate the Godbillon-Vey class in the case of some specific actions of groups on the circle. We will then show how these calculations lead to an example due to Thurston showing how one can vary the Godbillon-Vey "number."

Given an action of G on S^1, recall that on BG, ω is given by the 2 - cocycle

$$\omega(g,f) = \int_{S^1} \log(\mu_f)\, d\log(\mu_{gf}) \quad .$$

If G is a discrete subgroup of $PSL(2,\mathbb{R})$, then we have a natural action of G on S^1, which we view as the boundary of the upper half plane, given by the linear fractional transformations -

$$\begin{pmatrix} a & b \\ c & d \end{pmatrix}(z) = \frac{az+b}{cz+d}$$

For ease in computation, we can conjugate this action by the linear fractional transformation $z \to \frac{z-i}{z+i}$ taking the upper half plane into the disk $|z| \leq 1$ - we will want to pass back and forth freely between the two pictures.

Viewing f and g as acting on $|z| = 1$, we may now write

$$f(z) = \theta_1\left(\frac{z-\eta_1}{\bar{\eta}_1 z - 1}\right); \quad g \circ f(z) = \theta_2\left(\frac{z-\eta_2}{\bar{\eta}_2 z - 1}\right)$$

with $|\theta_1| = |\theta_2| = 1$, $|\eta_1|$, $|\eta_2| < 1$

and so viewing S^1 as $\mathbb{R}/\mathbb{Z}$, then f and g become

$$f(x) = \frac{1}{2\pi i}\log(\theta_1(\frac{e^{2\pi ix} - \eta_1}{\overline{\eta}_1 e^{2\pi ix} - 1})) \; ; \quad g\circ f(x) = \frac{1}{2\pi i}\log(\theta_2(\frac{e^{2\pi ix} - \eta_2}{\overline{\eta}_2 e^{2\pi ix} - 1})) \; .$$

Choosing the standard volume on S^1, $\mu_f = f'$ and $\mu_g = g'$, we can now easily compute by residues to get

$$\begin{aligned}\omega(f,g) &= \int_{S^1} \log(\mu_f)\, d\log(\mu_{gf}) \\ &= \int_0^1 \log(1 - e^{-2\pi ix}\eta_1)(1 - e^{2\pi ix}\overline{\eta}_1)[2\pi i]\left(\frac{e^{-2\pi ix}\eta_2}{1 - e^{-2\pi ix}\eta_2} - \frac{\overline{\eta}_2}{1 - e^{2\pi ix}\overline{\eta}_2}\right) dx \\ &= 2\pi i \log\left(\frac{1 - \eta_2\overline{\eta}_1}{1 - \overline{\eta}_2\eta_1}\right) \; .\end{aligned}$$

We can give this number a somewhat more geometric flavor by now passing to the upper half plane - writing $\eta_1 = \frac{a - i}{a + i}$, $\eta_2 = \frac{b - i}{b + i}$, we have

$$\omega(f,g) = 2\pi i \log\left(\frac{a+i}{\overline{a} - i}\right)\left(\frac{\overline{b} - i}{b + i}\right)\left(\frac{\overline{a} - b}{\overline{b} - a}\right) \; .$$

Now in the upper half plane, a geodesic is of the form $re^{i\theta} + k$, where r, k and θ are real. So given $z_1 = re^{i\theta_1} + k$, $z_2 = re^{i\theta_2} + k$, the expression $\log(\frac{z_1 - \overline{z}_2}{z_2 - \overline{z}_1}) = i(\theta_1 - \theta_2)$.

Hence well-known formulas in non-Euclidean geometry give us

$$(*) \qquad \omega(g,g) = (-4\pi^2)\ \text{area}\ (\Delta) \; ,$$

where Δ is the geodesic triangle with vertices i, a, and b. Here, of course, a and b satisfy $f(a) = i$, $(g\circ f)(b) = i$.

The formula $(*)$ relates well to the formulas of [5] in the following manner: If a Lie group G operates on a manifold M of dimension q, then we get a map from $\mathfrak{g} = \text{Lie}(G)$ to the Lie algebra of vector fields on M in the following way: given $X \in \mathfrak{g}$, then it defines a one-parameter group f_t of diffeomorphisms of M, and the infinitesimal generator of this flow is a vector field on M.

Let K be a maximal compact subgroup of G, with Lie algebra $\mathfrak{k}$, and choose a K-invariant volume θ on M. Then $\mathcal{L}_X(\theta) = \mu_X \cdot \theta$ defines a function μ_X on M, which we may think of as the "divergence" of X. One may define an "infinitesimal" Godbillon-Vey class by the formula

$$\omega(X_0, X_1, \cdots, X_q) = \int \mu_{X_0} \, d\mu_{X_1} \wedge \cdots \wedge d\mu_{X_q}$$

and one sees easily that this defines an element of $H^{q+1}(\mathfrak{g}/\mathfrak{k})$, independent of θ.

In [5] is constructed, given a discrete subgroup Γ of G, a map $H^*(\mathfrak{g}/\mathfrak{k}) \to H^*(B\Gamma)$. In the specific case of our standard $PSL(2, \mathbb{R})$ action on S^1, one can check easily that this "infinitesimal" Godbillon-Vey class agrees with our "global" Godbillon-Vey class.

Using the formula $(*)$, we can construct foliated circle bundles having arbitrary Godbillon-Vey number, according to the following scheme due to Thurston. Let a and b be any two points in the upper half plane. Then there is a unique parallelogram (in the sense that opposite sides are of equal length) having as three vertices i, a, and b. Labelling the fourth vertex c, then there are unique elements f and g satisfying

$$f(c) = a \qquad f(b) = i$$

$$g(c) = b \qquad g(a) = i$$

Then $fgf^{-1}g^{-1}(i) = i$, and so $fgf^{-1}g^{-1}$ is a rotation. A little non-Euclidean geometry shows us that the rotation is the non-Euclidean area of the parallelogram. Since the parallelogram may have an arbitrary area, we have shown

LEMMA. For any $0 < \alpha < 2\pi$, there are $f, g \in PSL(2, \mathbb{R})$ whose commutator is a rotation through angle α.

Now let G be the fundamental group of the two-holed torus. G is given as $\{X, Y, Z, W : XYX^{-1}Y^{-1} = ZWZ^{-1}W^{-1}\}$, and BG has a fundamental 2 - cycle

$$[BG] = (X, Y) - (XYX^{-1}, X) - (XYX^{-1}Y^{-1}, Y) - (Z, W) + (ZWZ^{-1}, Z) + (ZWZ^{-1}W^{-1}, W)$$

and this is the object on which, for careful choices of $X, Y, Z,$ and W, we want to evaluate our cocycle ω.

Fixing a positive integer n, and some $\alpha < \frac{2\pi}{n}$, choose f and g with commutator α, and f' and g' with commutator $n\alpha$.

Set $X = g$, and $Y = f$. We set $\widetilde{f} = (x) = \frac{1}{n} f'(nx)$, and similarly for $\widetilde{g}$ - we may think of $\widetilde{f}$ and $\widetilde{g}$ as liftings of f' and g' to an n - fold covering of the circle. Of course it is clear that the commutator of $\widetilde{f}$ and $\widetilde{g}$ is now α.

Now $\omega(g, f) - \omega(gfg^{-1}, g)$ is simply $(-4\pi^2) \cdot \alpha$, as can be seen by applying $(*)$ to the parallelogram with vertices i, a, b, and c. $\omega(gfg^{-1}f^{-1}, g) = 0$, since $gfg^{-1}f^{-1}$ is a rotation. Similarly, $\omega(g', g') - \omega(g'f'g'^{-1}, g') = (-4\pi^2) \cdot n\alpha$, and it is easy to check by the original integral formula that passing to an n - fold covering simply multiplies the result by n. Hence choosing $X = g$, $Y = f$, $Z = \widetilde{g}$, $W = \widetilde{f}$, we get $\omega([BG]) = 4\pi^2(n^2 - 1)\alpha$.

Varying α between 0 and $\frac{2\pi}{n}$, and taking n arbitrarily large, we see that ω may take any real value.

Bibliography

1. R. Bott, M. Mostow, J. Perchik, Gelfand-Fuks cohomology and foliations, Proceedings of the Eleventh Symposium New Mexico State University, 1973.

2. R. Bott, Some aspects of Invariant theory, C.I.M.E. Lectures delivered at Varenna, August 1975, to be published.

3. R. Bott, Lectures on characteristic classes and foliations (Notes by Lawrence Conlon), Lecture Notes in Mathematics vol. 279, 1-94. Springer-Verlag, New York.

4. R. Bott, H. Shulman, J. Stasheff, On the de Rham theory of certain classifying spaces, to be published in Advances of Math.

5. J. L. Dupont, Semisimplicial de Rham cohomology and characteristic classes of flat bundles, to be published in Top.

6. J. Heitsch, Independant variation of secondary classes, to be published.

7. F. W. Kamber, P. Tondeur, Foliated bundles and characteristic classes, Lecture Notes in Mathematics vol. 493, Springer-Verlag, New York.

8. W. Thurston, Foliations and groups of diffeomorphisms, Bull. AMS, 80, 304-312.

DE RHAM THEORY FOR $B\Gamma$

Herbert Shulman and James Stasheff

In our earlier paper with Bott [5], we studied the cohomology map $H^*(BG) \to H^*(B\Gamma)$ via a map of double complexes. The present paper is a sequel and will use much of the same notation.

In section 1 we identify the E_1 term of one of the spectral sequences associated to $\Omega^*(N\Gamma)$. We show in the second section how the double complex method leads naturally to characteristic classes of foliations. The third section describes how this approach gives insight into some properties of these classes.

§ 1. As in [5], Γ = germs of local diffeomorphisms of R^n, $\Gamma^{(p)}$ denotes composable p-tuples $\xleftarrow{\gamma_1} \cdot \ldots \cdot \xleftarrow{\gamma_p}$, while $\Omega^*(\Gamma^{(*)}) = \Omega^*(N\Gamma)$ is the double complex $\bigoplus_{p,q} \Omega^q(\Gamma^{(p)})$ with differentials $d_1 : \Omega^q(\Gamma^{(p)}) \to \Omega^{p+1}(\Gamma^{(p)})$ and $d_2 : \Omega^q(\Gamma^{(p)}) \to \Omega^q(\Gamma^{(p+1)})$. Filtering in the q-direction gives a spectral sequence with

$$E_1^{*,q} = H_{d_2}(\Omega^q(\Gamma^{(*)})) ,$$

the identification of which is the main result of this section.

First we identify the E_0 term :

LEMMA 1. <u>There is an isomorphism of</u> $\Omega^0(\Gamma^{(p)})$<u>-modules</u> : $\Omega^q(\Gamma^{(p)}) \approx C^\infty(\Gamma^{(p)}; \Lambda^q(R^{n^*}))$. (Here C^∞ denotes smooth maps and Λ^q the qth exterior power.)

<u>Proof</u> : Let $s : \Gamma^{(p)} \to R^n$ be the source map (of the composite, which is the source of the last coordinate). Since s is a local diffeomorphism, it pulls back the tangent bundle of R^n to the tangent bundle of $\Gamma^{(p)}$. A trivialization of TR^n will

then induce a trivialization of $T\Gamma^{(p)}$ and hence of $\Lambda^q(T^*\Gamma^{(p)})$, the sections of which are $\Omega^q(\Gamma^{(p)})$. The result follows from this.

Now we must describe what d_2 corresponds to in $C^\infty(\Gamma^{(p)};\Lambda^q(R^{n^*}))$ under the isomorphism. To do this, we introduce the smooth cohomology of a smooth pseudogroup C with coefficients in a smooth right C module V . This is the cohomology of the complex $\ldots\to C^\infty(C^{(p)};V) \xrightarrow{\delta} C^\infty(C^{(p+1)};V) \to\ldots$ where

$$\delta f(\gamma_1,\ldots,\gamma_{p+1}) = f(\gamma_2,\ldots,\gamma_{p+1}) +$$

$$\sum_{i=1}^{p} (-1)^i f(\ldots,\gamma_i\gamma_{i+1},\ldots) + (-1)^p f(\gamma_1,\ldots,\gamma_p)\cdot\gamma_{p+1}$$

and is denoted $H^*_{\text{smooth}}(C;V)$. In our case, we take $\Lambda^q(R^{n^*})$ as a smooth Γ-module via $\nu : \Gamma \to GL(n;R)$ and the standard action of $GL(n;R)$ on $\Lambda^q(R^{n^*})$.

THEOREM. $H_{d_2}(\Omega^q(\Gamma^{(*)})) \approx H^*_{\text{smooth}}(\Gamma;\Lambda^q(R^{n^*}))$.

Proof : Recall that $d_2 = \sum_{i=0}^{p+1} (-1)^i m_i^\#$ where

$$m_0(\gamma_1,\ldots,\gamma_{p+1}) = (\gamma_2,\ldots,\gamma_{p+1}) ,$$

$$m_{p+1}(\gamma_1,\ldots,\gamma_{p+1}) = (\gamma_1,\ldots,\gamma_p) ,$$

and

$$m_i(\gamma_1,\ldots,\gamma_{p+1}) = (\gamma_1,\ldots,\gamma_i\gamma_{i+1},\ldots,\gamma_{p+1}) , \quad 1 \le i \le p .$$

We must show that the isomorphism of Lemma 1 identifies δ with d_2 . In fact this occurs term by term. For $i < p+1$, we have $s\cdot m_i = s$ and so

$$\begin{array}{ccc} T\Gamma^{(p+1)} & \xrightarrow{Tm_i} & T\Gamma^{(p)} \\ \downarrow F & & \downarrow F \\ \Gamma^{(p+1)}\times R^n & \xrightarrow{m_i\times 1} & \Gamma^{(p)}\times R^n \end{array}$$

commutes, where F is the isomorphism of Lemma 1. For $i = p+1$ we have a more subtle situation : we must show

$$f(\gamma_2,\ldots,\gamma_{p+1})\cdot\gamma_{p+1} = m^{\#}_{p+1} f(\gamma_1,\ldots,\gamma_{p+1})$$

for f as above.

The crucial point in this comparison is the identification of the action $f.\gamma$ via the isomorphism F. We work with the left action in $T\Gamma^{(2)}$ first.

LEMMA. If $\tau \in T\Gamma^{(2)}$ and $F\tau = (\gamma_1,\gamma_2;v) \in \Gamma^{(2)} \times R^n$ then $F(m_{2*}\tau) = (\gamma_1;\gamma_2.v) \in \Gamma \times R^n$.

Proof : Let $\bar{\gamma}_1$ and $\bar{\gamma}_2$ be representative local diffeomorphisms of γ_1 and γ_2, i.e. γ_i is the germ of $\bar{\gamma}_i|_{x_i}$ where x_i = source (γ_i). A representative curve for τ in $\Gamma^{(2)}$ is then given by $(\bar{\gamma}_1|_{\bar{\gamma}_2(x_2+vt)}, \bar{\gamma}_2|_{x_2+vt})$, so that $m_{2*}\tau$ is represented by the curve $\bar{\gamma}_1|_{\bar{\gamma}_2(x_2+vt)}$. Thus $F(m_{2*}\tau) = (\gamma_1;w)$ where w is the tangent to the curve $\bar{\gamma}_2(x_2+vt)$ at $t=0$. This w is by definition $\gamma_2.v$.

The same analysis applies to $\Omega^q(\Gamma^{(p)})$ for all p and q, though for $p=0$ we have $F(m_{1*}\tau) = (s(\gamma_1);\gamma_1.v) \in \mathrm{Ob}\,\Gamma \times R^n = R^n \times R^n$. This proves the theorem.

Now consider the map $\nu : \Gamma \to G = GL(n;R)$ and the induced map $\nu^* : \Omega^*(G^*) \to \Omega^*(\Gamma^{(*)})$ of double complexes and of the corresponding spectral sequences. In [3], Bott showed that for $\Omega^*(G^*)$,

$$E_1^{p,q} \approx H_c^{p,q}(G;S^q(g^*)) .$$

This is the continuous Eilenberg-Mac Lane cohomology, where g is the Lie algebra, S^q is the qth symmetric power and G acts by the adjoint action. The map of E_1 terms

$$H_c^{p-q}(G,S^q(g^*)) \approx H^{p-q}_{smooth}(G;S^q(g^*)) \to H^p_{smooth}(\Gamma;\Lambda^q(R^{n^*}))$$

has, at present, no direct description (e.g. at the chain level).

§ 2. Since $\Omega^q(\Gamma^{(*)}) = 0$ for $q>n$, the map $\nu^* : \Omega^*(G^*) \to \Omega^*(\Gamma^{(*)})$ induces $\nu^* : \Omega^*(G^*)/\mathfrak{I}_n \to \Omega^*(\Gamma^{(*)})$ where $\mathfrak{I}_n = \bigoplus_{q>n} \Omega^q(G^*)$ and is denoted $\mathfrak{I}_n\Omega^*NG$ in [5].

Elements of $H(\Omega^*(G^*)/\mathfrak{I}_n)$ are then characteristic classes for codimension n smooth foliations. The remainder of this section is devoted to computing $H(\Omega^*(G^*)/\mathfrak{I}_n)$.

First we need an explicit description of the Bott E_1 term, $H_c^{p-q}(G;S^q(g^*))$. By the van Est isomorphism, it is isomorphic to $H^{p-q}_{\text{Lie alg}}(g,K;S^q(g^*))$ where K is a maximal compact subgroup. Now $S^q(g^*)$ is a semi-simple G-module, with invariant elements denoted I_G^*, and a reasonably well-known procedure (see for example [7]) proves

$$H^*(g,K;S^q(g^*)) \approx H^*(g,K;I_G^q) .$$

This last module is seen to be

$$H^*(g,K;R) \otimes I_G^q \quad \text{for } G,K \text{ a reductive pair.}$$

(The purely Lie algebraic spectral sequence of this form is due to Kamber and Tondeur [10 and related papers]. Their method allows the identification of Bott's spectral sequence with theirs precisely via the van Est isomorphism as we shall see.)

In particular, for $G = GL(n;R)$, it is known that $H^*(g,K;R) \approx H^*(U(n)/K;R) = E[h_1,h_3,\dots,h_{2[\frac{n+1}{2}]-1}]$ where E denotes exterior algebra and $\deg h_i = 2i-1$. Also, $I_G^* = R[c_1,\dots,c_n]$ with $\deg c_i = 2i$.

Now the spectral sequence converges to

$$H^*(BG) = R[c_2,c_4,\dots,c_{2[\frac{n}{2}]}] .$$

It follows that d_1 (of the spectral sequence) sends h_1 to a non-zero multiple of c_1. Proceeding by induction, we see that c_{2i-1} and h_{2i-1} survive to E_{2i-1} and at that stage, d_{2i-1} must send h_{2i-1} to a multiple of c_{2i-1} because the odd c_j's of degree less than $2i-1$ have already been killed by the E_{2i-1} stage (i.e., the indeterminacy is zero for dimension reasons). Note, too, that $d_r = 0$ for $r > 2[\frac{n+1}{2}]$.

The above facts are also true for the analogous spectral sequence of $\Omega^*(G^*)/\mathfrak{I}_n$. The difference shows up as a truncation of $R[c_1,\dots,c_{2[\frac{n}{2}]}]$ and in the appearance of new cocycles. To compute the cohomology of $\Omega^* G^* / \mathfrak{I}_n$, we employ a certain amount of differential algebra which is designed for the study of characteristic classes [6]. In considering the de Rham cohomology of a bundle, the essential information is contained in the connection. The universal algebra in this context is the Weil algebra $W(g)$ whose definition we now recall. As an algebra, it is the tensor product $\Lambda(\bar{g}) \otimes S(\bar{g})$ with the copy of $\bar{g}$, the dual of the Lie algebra of G , generating $S(\bar{g})$ considered to have degree 2 . For $g = g\ell_n$, let $w^i_j \in \Lambda^1(\bar{g})$ (respectively $\Omega^i_j \in S^1(\bar{g})$) be the dual bases to the canonical one in $g\ell_n$.

The differential in the algebra $W(g\ell_n)$ is determined by $dw^i_j = -w^i_k \wedge w^k_j + \Omega^i_j$ and $d^2 = 0$. This makes $W(g\ell_n)$ an <u>acyclic</u> differential $g\ell_n$-algebra, i.e. d is compatible with the adjoint representation (Lie derivative) and interior product (evaluation on elements of $g\ell_n$). We abbreviate $g\ell_n$ to g henceforth. We regard $W(g)$ as filtered by the ideals $\mathfrak{I}_r$ generated by $\bigoplus_{q>r} S^q(\bar{g})$.

We further denote by $W(g,K)$ the subalgebra of K-basic elements , which inherits the filtration $\mathfrak{I}_r$.

Now the cohomology $H^*(\Lambda^*(\bar{g}))$ can be computed as $E(h_1,\ldots,h_n)$, h_i of degree $2i-1$, and in fact identified as a deformation retract of $\Lambda^*(\bar{g})$, namely the exterior algebra of biinvariant forms. Similarly the invariant polynomials $R[c_1,\ldots,c_n] = I_G \subset S(\bar{g})$ pick out the cohomology. Indeed the differential in $W(g)$ restricts to $du_i = c_i$ so that $E(u_i) \otimes R[c_i]$ is acyclic.

Now we denote by WO_n the quotient $E(u_1,u_3,\ldots,u_{2[\frac{n}{2}]+1}) \otimes R[c_1,\ldots,c_n]/\mathfrak{I}_n$.

THEOREM. [10, 14]. <u>The inclusion</u> $WO_n \subset W(g,O_n)/\mathfrak{I}_n$ <u>induces an isomorphism in cohomology.</u>

More interesting than the statement is the proof. Both complexes are filtered and one compares the spectral sequences. The E_1-term for WO_n is just WO_n while for $W(g,O_n)/\mathfrak{I}_n$ we have a truncated version of that of Hochschild-Serre, i.e. $H(g\ell_n,O_n) \otimes R[c_1,\ldots,c_n]/\mathfrak{I}_n$. The equivalence $E(h_1,h_3,\ldots,h_{2[\frac{n}{2}]+1}) \approx H(g\ell_n,O_n)$ gives the desired comparison. Indeed this is isomorphic to the E_1-term for $\Omega^* G^* / \mathfrak{I}_n$ and in fact we have such precise control over the differentials in both spectral sequences that we can <u>formally</u> deduce at least the additive isomorphism :

$$H(\Omega^* G^*/\mathfrak{F}_n) \approx H(WO_n) .$$

However, it is somewhat more satisfying to be able to compare $WO_n \subset W(g,K)/\mathfrak{F}_n$ directly with $\Omega^* G^*/\mathfrak{F}_n$. This can be done in a variety of ways, all involving an auxiliary differential algebra.

The Weil algebra $W(g)$ is of interest precisely because it is universal in the sense that given any principal $G = GL(n,R)$-bundle $E \to B$ with connection $w : \bar{g} \to \Omega^*(E)$, there is a map of g-DG-algebras $W(g) \to \Omega^*(E)$. The image $\pi^* \Omega^*(B)$ in $\Omega^*(E)$ can be identified as the g-basic forms.

To compare $W(g)/\mathfrak{F}_n$ with $\Omega^*(G^*)/\mathfrak{F}_n$ we need either to talk about $\Omega^*(EG)$ in some meaningful way or to have an analogue of $W(g)$ in the corresponding simplicial category. The latter is more appropriate in the present context and has been well worked out by Kamber and Tondeur [10]. They define a simplicial g-DG-algebra

$$W_1(g) = \bigoplus_p W(g^p)$$

with faces induced by the projections $\pi_i : G^p \to G^{p-1}$, omitting the ith factor. Projection on the summand $p=1$ gives a canonical map of g-DG-algebras $W_1(g) \to W(g)$ which is a chain equivalence. Indeed they show more : Let $W_1(g,K)$ denote the K-basic elements of $W_1(g)$. For a suitable filtration $\mathfrak{F}_n$ using the associated spectral sequence, Kamber and Tondeur show :

THEOREM. [2.10, 10a]. <u>The projection</u> $W_1(g,K)/\mathfrak{F}_n \to W(g,K)/\mathfrak{F}_n$ <u>induces an isomorphism in cohomology.</u>

In particular, this gives the analog of our proposition :

$$H(W_1(g,K)/\mathfrak{F}_n) \approx H(WO_n) .$$

On the other hand, consider the universal bundle analog of our complex $\Omega^*(G^*)$, i.e. the homogeneous complex $\bigoplus_{p \geq 0} \Omega^*(G^{p+1})$ with faces induced by the projections π_i and the diagonal action of g. By abuse of language we call this $\Omega^*(EG)$. It can be identified with the non-homogeneous complex $\bigoplus_{p \geq 0} \Omega^*(G^{p+1})$

with faces induced by the maps $m_i : (g_1,\dots,g_{p+1}) \to (\dots,g_ig_{i+1},\dots)$ and the action of g on the last coordinate. (Send $g_0,\dots,g_{p+1}$ to $g_0\cdots g_{p+1},\dots,g_pg_{p+1},g_{p+1}$.) Our complex $\Omega^*(G^*)$ can be identified with the g-basic forms. Ths obvious flat connections on $G^{p+1} \to *$ give $W(g^{p+1}) \overset{\varphi}{\to} \Omega^*(G^{p+1})$ forming a map of simplicial g-DG-algebras. To filter $\Omega^*(EG)$ compatibly with Kamber and Tondeur, we define $\mathfrak{F}_n(\Omega^*(G^{p+1}))$ as follows :

Let F_n be the filtration by form degree on the G-basic complex, i.e.

$$F_n = \bigoplus_{q>n} \Omega^q(G^*) .$$

Then $\mathfrak{F}_n$ is the filtration on $\Omega^*(EG)$ induced from F_n by $\Psi^\# : \Omega^*(G^p) \to \Omega^*(G^{p+1})$ where $\Psi : G^{p+1} \to G^p$ is the projection on the first p coordinates.

$\mathfrak{F}_n$ also restricts to a filtration on $\Omega^*(EG)_{K\text{-basic}}$, and $\Psi^\# : \Omega^*(EG)_{G\text{-basic}} \to \Omega^*(EG)_{K\text{-basic}}$ is filtration preserving. We then obtain the following key result which can be thought of as a generalized Van Est isomorphism :

PROPOSITION. $\Psi^\#$ <u>induces an isomorphism of spectral sequences.</u>

<u>Proof</u> : The above comparison of the homogeneous and non-homogeneous simplicial spaces for EG shows that $\Omega^*(EG)_{K\text{-basic}} \approx \bigoplus_p \Omega^*(G^p;\Omega^*(G/K))$ and that $\mathfrak{F}_n = \bigoplus_{p,\, q>n} \Omega^q(G^p;\Omega^*(G/K))$. Then

$$\mathfrak{G}_n = \mathfrak{F}_{n-1}/\mathfrak{F}_n = \bigoplus_p \Omega^n(G^p;\Omega^*(G/K))$$

and $d_o : \mathfrak{G}_n \to \mathfrak{G}_n$ is given by $d_o = \delta + d_{G/K}$ where

$$\delta : \Omega^n(G^p;\Omega^*(G/K)) \to \Omega^n(G^{p+1};\Omega^*(G/K))$$

is the simplicial coboundary and

$$d_{G/K} : \Omega^r(G/K) \to \Omega^{r+1}(G/K)$$

is the de Rham differential.

Note that $\mathfrak{G}_0$ is precisely the double complex used by Van Est in the proof of his isomorphism, and the maps $\Psi_o^\#$ and $\varphi_o^\#$ below are precisely the maps used [13] .

Thus $(\mathcal{G}_n, d_o)$ is a double complex and its homology $H_{d_o}(\mathcal{G}_n)$ can be computed via a spectral sequence with

$$E_2 = H_\delta H_{d_{G/K}}(\mathcal{G}_n) \Rightarrow H_{d_o}(\mathcal{G}_n) .$$

Now $H_{d_{G/K}}(\mathcal{G}_n) \approx \Omega^n(G^P;R)$ since $G/K \simeq *$, and this gives

$$H^P_{d_o}(G_n) \approx H_\delta(\Omega^n(G^P;R)) .$$

For the G-basic complex we have

$$G'_n = F_{n-1}/F_n = \bigoplus_p \Omega^n(G^P;R) \quad \text{with} \quad d'_o = \delta$$

and it follows that

$$\Psi^{\#} : (G'_n, d'_o) \to (\mathcal{G}_n, d_o)$$

is a chain equivalence and that $\Psi^{\#}$ induces an isomorphism of spectral sequences.

In a similar way it is useful to compare the above spectral sequence for $\Omega^*(EG)_{K\text{-basic}}$ to the one for $W_1(g;k)$ obtained by Kamber-Tondeur :

PROPOSITION. <u>The map</u> φ <u>above induces a filtration preserving map</u> $\varphi^{\#} : W_1(g;k) \to \Omega^*(EG)_{K\text{-basic}}$ <u>and an isomorphism of spectral sequences.</u>

<u>Proof</u> : Elements of the nth filtration $\bar{F}_n$ in $W_1(g;k)$ represent forms of degree greater than n , so that $\varphi^{\#}$ is filtration preserving. For $n = 0$, we have

$$\bar{G}_0 = W_1(g;k)/F_1 = \bigoplus_p \Lambda^*(g^P)_{K\text{-basic}}$$

and the map $\bar{G}_0 \xrightarrow{\varphi_o^{\#}} \mathcal{G}_0$ is the one induced by $\Lambda(g)_{K\text{-basic}} \to \Omega^*(G/K)$. As in the proof of the Van Est isomorphism, $\varphi_o^{\#}$ is an isomorphism.

We also have that $\varphi^* : H(W_1(g;k)) \to H(\Omega^*(EG)_{K\text{-basic}})$ is an isomorphism since $W_1(g)$ and $\Omega^*(EG)$ are acyclic and φ^* is k-equivariant. Also the calculations of Bott and Kamber-Tondeur for the spectral sequences for $\Omega^*(EG)_{K\text{-basic}}$ and $W_1(g;k)$ respectively show that the E_1-terms are tensor products.

The standard spectral sequence comparison theorem then gives the desired result.

We are now able to calculate the maps of truncated complexes :

$$\begin{array}{ccc} & & W_1(g;k)/\bar{F}_n \\ & & \big\downarrow \varphi_n^{\#} \\ \Omega^*(EG)_{G\text{-basic}/F_n} & \xrightarrow{\Psi_n^{\#}} & \Omega^*(EG)_{K\text{-basic}/\mathfrak{F}_n} \end{array} .$$

We still get maps of spectral sequences as above, the only difference being that the E_0-terms are truncated, i.e.

$$E_0^{p,q} = 0 \quad \text{for} \quad q > n .$$

The maps for E_1 are still isomorphisms and thus $\varphi_n^{\#}$ and $\Psi_n^{\#}$ induces isomorphisms on homology.

Combining this fact with the previous two theorems finally gives the natural isomorphism

$$H(WO_n) \approx H(\Omega^*(G^*)/F_n) ,$$

as algebras, in fact respecting all the structure of higher order (matrix) Massey products, which in this case are known all to be zero in $H(WO_n)$. The composition

$$H(WO_n) \approx H(\Omega^*(G^*)/\mathfrak{F}_n) \xrightarrow{\bar{\nu}^*} H(\Omega^*(\Gamma^*)) \longrightarrow H(B\Gamma)$$

gives an alternative description of the "linear" characteristic classes for foliations, defined by a number of others [1, 4, 8, 10]. Of these, the method of Kamber and Tondeur is closest to our own.

A pious hope is that the composite is a monomorphism. It is at least non-zero on the E_1 level. Consider the edge $H_c(G)$. From the work of Borel and Selberg [2], it is known that for $G = GL_n(R)$ we have a monomorphism $H_c(G) \to H(BG^{\delta})$, the cohomology of the discrete group $GL_n(R)$. From the sequence $G^{\delta} \to \Gamma \to G$, we see that $H_c(G) \to H_c(\Gamma)$ is also monic.

From the known $(n+1)$-connectedness of $B\Gamma_n \to BGL_n(R)$, one can deduce that $d_1 h_1 = c_1 \neq 0$ in the spectral sequence for Γ_n. (The hoped for

2n-connectedness would imply $d_n h_{2n-1} = c_n \neq 0$.)

On the other edge, examples of Thurston [12] show $h_1 c_1^n \neq 0$ in $H^{2n+1}(B\Gamma_n)$. Since $E_1^{p,q}(\Gamma_n) = 0$ for $q > n$, $0 \neq h_1 c_1^n \in E_1^{n+1,n}$. (For the oriented version of Γ , a similar argument applies to χ^2 , the square of the Euler class.)

The naturality of our isomorphism $H(WO_n) \approx H(\Omega^* G^*/F_n)$ (revealed by the Kamber and Tondeur approach) also leads to a simplification of Morita's result [11]: The classes $h_1 c_1^r c_2^s$ for $r+2s = n$ are independent in $H^{2n+1}(B\Gamma_n)$, $n \geq 2$. Other techniques of Kamber and Tondeur [10c] show the classes $h_1 h_{i_1} \dots h_{i_s} c_1^{j_1} \dots c_n^{j_n}$ for $2 \leq i_1 < \dots < i_s \leq \left[\frac{n+1}{2}\right]$ and $\deg c_J = 2n$ are independent.

§ 3. The Heitsch rigidity theorem [9] describes the invariance of certain WO_n-classes in $B\Gamma_n$ under homotopy of foliations (i.e. continuous deformations). As explained in [9], the classes in the image of $H^*(WO_{n+1}) \to H^*(WO_n)$ are rigid. (See also Theorem 8.9 of [10].) More generally, the image of $r^* : H^*(B\Gamma_{n+1}) \to H^*(B\Gamma_n)$ consists of rigid characteristic classes of foliations (r^* being induced from the standard inclusion $R^n \hookrightarrow R^{n+1}$).

For the double complexes, we have $\Omega^*(\Gamma_{n+1}^{(*)}) \to \Omega^*(\Gamma_n^{(*)})$ which induces a map of E_1-terms

$$H^*_{sm}(\Gamma_{n+1}; \Lambda^q(R^{n+1*})) \to H^*_{sm}(\Gamma_n; \Lambda^q(R^{n*}))$$

which we denote by $E_1(n+1) \xrightarrow{r^*} E_1(n)$. To obtain non-rigid elements we proceed as follows : Consider an element $x \in E_1(n+1)$ which does not live to $E_\infty(n+1)$. If $r^* x \in E_1(n)$ does live to $E_\infty(n)$, then it will represent a class in coker(r^*) . One such choice is an x such that $d_{i-1} x = 0$ but $d_i x = y \neq 0$, where the d_i are the differentials in $E_i(n+1)$ and $x \in E_1^{p,n-i+1}$ so that $y \in E_1^{p-i+1,n+1}$ and hence $r^* y = 0$.

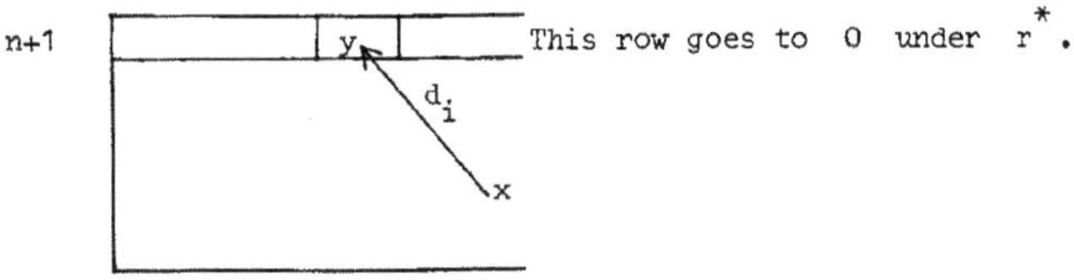

This line of reasoning applied to the image of WO_n leads to at least all the non-rigid classes Heitsch describes in [9].

We have chosen to present all our arguments in terms of general foliations. Analogous results can be worked out for special cases of interest : complex, Hamiltonian, etc.

BIBLIOGRAPHY

[1] I.N. Bernstein and B.I. Rozenfeld, On characteristic classes of foliations, Funk. Anal. i Pril 6 (1972), 68-69.

[2] A. Borel, Compact Clifford-Klein forms of symmetric spaces, Topology 2 (1963), 111-122.

[3] R. Bott, On the Chern-Weil homomorphism and the continuous cohomology of Lie groups, Advances in Math. 11 (1973), 289-303.

[4] R. Bott and A. Haefliger, On characteristic classes of Γ-foliations, Bull. A.M.S. 78 (1972), 1039-1044.

[5] R. Bott, H. Shulman and J. Stasheff, On the de Rham theory of classifying spaces, to appear in Advances in Math.

[6] H. Cartan, Notions d'algèbre différentielle, etc..., Colloque de Topologie, Bruxelles (1950), 15-27 and 57-71.

[7] W. Greub, S. Halperin and R. Van Stone, Connections, Curvature and Cohomology, Vol. III, Acad. Press 1976.

[8] A. Haefliger, Sur les classes caractéristiques des feuilletages, Sém. Bourbaki 1971/72, # 412, Springer Lecture Notes in Math., 317 (1973).

[9] J. Heitsch, Deformations of secondary characteristic classes, Topology 12 (1973), 381-388.

[10a] F. Kamber and P. Tondeur, Characteristic invariants of foliated bundles, Manuscripta Mathematica 11 (1974), 51-89.

[10b] F. Kamber and P. Tondeur, Semi-simplicial Weil algebras and characteristic classes for foliated bundles in Cech cohomology, Proc. Symposia Pure Math., Vol. 27, 283-294.

[10c] F. Kamber and P. Tondeur, Foliated Bundles and Characteristic Classes, Springer Lecture Notes in Math., n° 493 (1975).

[11] S. Morita, A remark on the continuous variation of secondary characteristic classes for foliations, I.A.S. preprint.

[12] W. Thurston, Variations of the Godbillon-Vey invariant in higher codimensions, to appear.

[13] Van Est, Une application d'une méthode de Cartan-Leray, Indag. Math. 17 (1955), 542-4.

[14] C. Godbillon, Cohomologies d'algèbres de Lie de champs de vecteurs formels, Séminaire Bourbaki 1972/73, exposé 421, Springer Lecture Notes in Math., n° 383 (1974).

Differential Geometry and Foliations: The Godbillon-Vey Invariant and the Bott-Pasternack Vanishing-Theorems

by

Robert B. Gardner

<u>Definition.</u> A C^r-foliation of codimension p on m-dimensional manifold M is a decomposition of M into a union of disjoint connected subsets $\{L_i\}_{i\in I}$ called leaves of the foliation. This decomposition has the property that every point in M has a neighborhood U and a system of local class C^r-coordinates $(u^1,\ldots,u^m) : U \to R^m$ such that for each leaf L_i, the components of $U \cap L_i$ are described by the equations

$$u^{m-p+1} = \text{constant},\ldots,u^m = \text{constant} \quad .$$

These neighborhoods U are called distinguished neighborhoods.

Using elementary topology one can extract a locally finite cover $\{U_\alpha,u_\alpha\}_{\alpha\in J}$ of distinguished neighborhoods and construct a partition of unity $\{\lambda_\alpha\}_{\alpha\in J}$ subordinate to the cover. Thus the λ_α are smooth functions having the properties that

$$\text{support } \lambda_\alpha \subset U_\alpha, \text{ and } 0 \le \lambda_\alpha \le 1$$

and

$$\sum \lambda_\alpha \equiv 1 \quad .$$

If (U,u) and (V,v) are two distinguished neighborhoods, then on $U \cap V$

$$dV = dU\gamma_{UV} \quad \text{where} \quad (\gamma_{UV})_{ij} = \frac{\partial v^j}{\partial u^i} \quad .$$

Invited survey address at Escola de Topologia, Pontifica Universidade Catolica, Rio de Janeiro, Brasil, January 5-24, 1976. This work was partially supported by the National Science Foundation under Grant MPS73-08685-A02.

Since the leaves of the foliation are independent of the coordinate chart, the differentials $\{dv^{m-p+1},\ldots,dv^m\}$ are linear combinations of the differentials $\{du^{m-p+1},\ldots,du^m\}$. Thus the matrix γ_{UV} has the form

$$\gamma_{UV} = \begin{pmatrix} * & 0 \\ * & g_{UV} \end{pmatrix} .$$

If the covering by distinguished neighborhoods can be so chosen that whenevery $U_\alpha \cap U_\beta \neq \emptyset$,

$$\det g_{U_\alpha U_\beta} > 0$$

we say the foliation is <u>transversally oriented</u>.

In this case we may introduce a global p-form Ω by defining

$$\begin{aligned} \Omega|_{U_\alpha} &= \sum_\beta \lambda_\beta du_\beta^{m-p+1}\wedge\ldots\wedge du_\beta^m \\ &= \sum_\beta \lambda_\beta \det g_{U_\beta U_\alpha} du_\alpha^{m-p+1}\wedge\ldots\wedge du_\alpha^m \ . \end{aligned}$$

Let I be the ideal of forms which vanish on every leaf, the restriction of I to a distinguished neighborhood U_α has generators

$$I|_{U_\alpha} = \{du_\alpha^{m-p+1},\ldots,du_\alpha^m\}$$

and since $\sum\lambda_\beta \det g_{U_\alpha U_\beta} > 0$ the 1-forms in this ideal are characterized by

$$\{\tau | \tau \wedge \Omega = 0\} \ .$$

A geometric consequence of the Frobenius theorem is that the maximal connected integral submanifolds of $\Omega = 0$ on U_α are precisely $U_\alpha \cap L_i$.

<u>Lemma</u>. If σ and η are 1-forms such that

$$d\sigma \wedge \Omega = 0 \qquad\qquad \eta \wedge \Omega = 0$$

then

$$(1) \qquad (d\sigma)^{p+1} = 0 \qquad \text{and} \qquad (2) \qquad (d\sigma)^p \wedge \eta = 0 \quad .$$

Proof. Locally on a distinguished neighborhood

$$d\sigma = \sum_{i=m-p+1}^{m} du_\alpha^i \wedge \theta^i \qquad \text{and} \qquad \eta = \sum_{i=m-p+1}^{m} a_i du_\alpha^i$$

hence (1) and (2) follow by linear dependence.

Let $\mu_\alpha = \sum \lambda_\beta \det g_{U_\alpha U_\beta}$, then the exterior derivative of Ω and U becomes

$$\begin{aligned} d\Omega &= d\mu_\alpha \wedge du_\alpha^{m-p+1} \wedge \dots \wedge du_\alpha^m \\ &= d \log \mu_\alpha \wedge \Omega \ , \end{aligned}$$

again using the partition of unity

$$d\Omega = \sum \lambda_\alpha d\Omega = (\sum \lambda_\alpha d \log \mu_\alpha) \wedge \Omega.$$

We see that

$$\omega = \sum \lambda_\alpha d \log \mu_\alpha$$

is a globally defined 1-form with the property that globally

$$d\Omega = \omega \wedge \Omega \quad .$$

Conversely given a global p-form Ω which is locally decomposable, that is

$$\Omega|_{U_\alpha} = \omega_\alpha^1 \wedge \dots \wedge \omega_\alpha^p \quad ,$$

and satisfies

$$d\Omega = \omega \wedge \Omega \quad ,$$

then the Frobenius theorem implies that there is a unique maximal connected integral submanifold of dimension $(m - p)$ through every point which are the leaves of a foliation.

On a distinguished neighborhood U_α

$$\Omega|_{U_\alpha} = f_\alpha du_\alpha^{m-p+1} \wedge \ldots \wedge du_\alpha^m$$

with $f_\alpha \neq 0$, and by changing u_α^{m-p+1} by a sign if necessary we can assume $f_\alpha > 0$.

As a result on any overlaping distinguished neighborhoods $U_\alpha \cap U_\beta \neq \emptyset$

$$\begin{aligned} \Omega|_{U_\alpha \cap U_\beta} &= f_\beta du_\beta^{m-p+1} \wedge \ldots \wedge du_\beta^m \\ &= f_\beta \det g_{\beta\alpha} du_\alpha^{m-p+1} \wedge \ldots \wedge du_\alpha^m \end{aligned}$$

and

$$f_\alpha = f_\beta \det g_{\alpha\beta} \quad \text{which implies} \quad \det g_{\beta\alpha} > 0$$

and hence that the foliation is transversely oriented. Further important properties of the Godbillon-Vey invariant can be found in [8].

§1. The Godbillon-Vey Invariant

We have seen that a transversally oriented foliation is equivalent to a global p-form Ω defined up to multiplication by a non-zero function which satisfies a relation

$$d\Omega = \omega \wedge \Omega.$$

Differentiation of this relation gives

$$\begin{aligned} 0 = d(d\Omega) &= d\omega \wedge \Omega - \omega \wedge d\Omega \\ &= d\omega \wedge \Omega - \omega \wedge (\omega \wedge \Omega) \\ &= d\omega \wedge \Omega \end{aligned}$$

and hence $d\omega$ is contained in the ideal I of forms vanishing on every leaf. This implies by (1) of the lemma in §0 that

$$(d\omega)^{p+1} \equiv 0 \quad .$$

In particular the $(2p+1)$-form $d\omega^p \wedge \omega$ is closed since

$$d(d\omega^p \wedge \omega) = d\omega^{p+1} = 0 \ .$$

Since neither Ω nor ω is uniquely defined by the foliation it is natural to study the effect of the different choices involved on the cohomology class of this closed $(2p+1)$-form.

If $\Omega' = \lambda\Omega$ with $\lambda \neq 0$ then

$$\begin{aligned} d\Omega' &= d\lambda \wedge \Omega + \lambda d\Omega = d\lambda \wedge \Omega + \lambda\omega \wedge \Omega \\ &= (\frac{d\lambda}{\lambda} + \omega)\wedge\Omega' = (d \log|\lambda| + \omega)\wedge\Omega' \end{aligned}$$

and

$$\begin{aligned} d(d \log|\lambda| + \omega)^p \wedge (d \log|\lambda| + \omega) &= (d\omega^p \wedge d \log|\lambda|) + d\omega^p \wedge \omega \\ &= d\omega^p \wedge \omega + d(\omega^p \wedge d \log|\lambda|) \ . \end{aligned}$$

If $d\Omega = \omega' \wedge \Omega$ then

$$0 = (\omega - \omega') \wedge \Omega$$

and $\omega - \omega'$ is contained in the ideal I of 1-forms which vanishes on every leaf. Thus

$$\omega' = \omega + \eta \quad \text{where} \quad \eta \wedge \Omega = 0 \ .$$

Now

$$d\omega'^p \wedge \omega' = (d\omega + d\eta)^p \wedge \omega + (d\omega + d\eta)^p \wedge \eta$$

and the second term is zero by (2) of the lemma in §0). Since

$$(d\omega + d\eta)^p \wedge \omega = d\omega^p \wedge \omega + \sum_{q=1}^{p} \binom{p}{q} d\omega^{p-q} \wedge d\eta^q \wedge \omega$$

$$= d\omega^p \wedge \omega + d\Big(\sum_{q=1}^{p} \binom{p}{q} d\omega^{p-q} d\eta^{q-1} \wedge \eta \wedge \omega\Big)$$

$$- \sum \binom{p}{q} d\omega^{p-q+1} \wedge d\eta^{q-1} \wedge \eta \ ,$$

and each term of the last sum again vanishes by (2) of the lemma in §0 we see that the de Rham class of $d\omega^p \wedge \omega$ is an invariant of the foliation.

We will write this class called the Godbillon-Vey class [3] as

$$gv(\Omega) = [d\omega^p \wedge \omega] \in H^{2p+1}(M;\mathbb{R}) \ .$$

A basic fact and an easy calculation is that if

$$\Omega = \Omega' \wedge \Omega'' \quad \text{with} \quad d\Omega' = \omega' \wedge \Omega' \quad \text{and} \quad d\Omega'' = \omega'' \wedge \Omega''$$

globally then

$$gv(\Omega) = [0]$$

see Sondow [5]. As a result a non-trivial class must have an irreducibility property.

The classes are genuinely non-trivial since W. Thurston has announced the following important theorem.

<u>Theorem</u>: Given any real number $r \in \mathbb{R}$ there exists a foliation of co-dimension p which is transversely orientable on a manifold of dimension $2p + 1$ such that

$$\int_M gv(\Omega) = r \quad .$$

The special case $p = 1$ has appeared in Thurston [7] in which he describes the invariant as a measure of helical wobble. This description is given more precise substance in Reinhart and Wood [6].

§2. Connections and Characteristic Classes

The notion of connection was developed by Christoffel in about 1870 in order to define a differentiation mapping tensor fields into tensor fields.

If x is a vector field locally defined on a coordinate neighborhood U by

$$x|_U = \sum x_U^i \frac{\partial}{\partial u^i} = x_U \frac{\partial}{\partial u}$$

and on a neighborhood V by

$$x|_V = x_V \frac{\partial}{\partial v}$$

then on $U \cap V$

$$x_V = x_U \gamma_{UV} \quad \text{where} \quad (\gamma_{UV})_{ij} = \frac{\partial v^j}{\partial u^i}.$$

As a result taking the differential of the components gives

$$dx_V = dx_U \gamma_{UV} + x_U d\,\gamma_{UV}$$

which no longer transforms like a tensor field. In particular the vanishing of the differentials of the components is not intrinsic.

One of the simplest modifications of this last operator is to add a linear term

$$Dx|_U = dx_U + x_U \theta_U$$

where θ_U is some matrix of 1-forms. In order that

$$Dx|_V = Dx|_U \gamma_{UV}$$

we see

$$dx_V + x_V \theta_V = dx_U \gamma_{UV} + x_U d\gamma_{UV} + x_U \gamma_{UV} \theta_V$$

must equal

$$dx_U\gamma_{UV} + x_U\theta_U\gamma_{UV}$$

and this for all vector field components x_U. Hence we must have

$$(1) \qquad d\gamma_{UV} + \gamma_{UV}\theta_V = \theta_U\gamma_{UV} \quad .$$

Since these conditions on the matrices of 1 forms θ_U only depend on the transition functions γ_{UV} it is natural to consider the general situation of a principal G-bundle.

$$\begin{array}{ccc} G & \longrightarrow & B_G \\ & & \downarrow \\ & & M \end{array} \quad .$$

If B_G defines transition functions $\{\gamma_{UV}\}$ relative to an open cover and if $\{\theta_U\}$ is a collection of matrices defined over the open sets of the cover and satisfy (1) then the collection is called a <u>connection on B_G</u>. This object gives rise to a reasonable process of differentiation of <u>$B_G x_\rho A$</u>, the sections of any associated bundle $B_G x_\rho A$ where

$$\rho : G \longrightarrow \text{Aut } A$$

is a representation.

Let

$$\phi_U : U \times A \longrightarrow B_G x_\rho A$$

be a strip map, then any section over U has local components σ_U defined by

$$\sigma(p) = \phi_U(p,\sigma_U(p))$$

As such we may define differentiation locally by

$$D\sigma|_U = d\sigma_U + \sigma_U\, \rho_*(\theta_U)$$

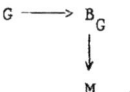

where

$$\rho_* : L(G) \longrightarrow \mathrm{Hom}(V,V)$$

is the induced mapping on the tangent space at the identity. As a result,

$$D\sigma|_V = D\sigma|_U \, \rho(\gamma_{UV})$$

and

$$D : \underline{B_G x_\rho V} \longrightarrow \underline{B_G x_\rho V} \otimes \underline{T^*(M)} .$$

This operation may be extended to tensorial p-forms by

$$D : \underline{B_G x_\rho V} \otimes \underline{\Lambda^p T^*(M)} \longrightarrow \underline{B_G x_\rho V} \otimes \underline{\Lambda^{p+1} T^*(M)}$$

where

$$D(\sigma_U \otimes \omega) = D\sigma_U \wedge \omega + \sigma_U \otimes d\omega .$$

A section $\sigma \in \underline{B_G x_\rho V}$ is called parallel if

$$D\sigma = 0 .$$

Example 1: If $\sigma_U = (du^1, \ldots, du^m)$ is a vector 1-form i.e. $\sigma(p) = \sum \frac{\partial}{\partial u^i} \otimes du^i$ then $\sigma_V = \sigma_U \gamma_{UV}$ and if $\theta_j^i = \sum \Gamma_j{}^i{}_k du^k$ then

$$D\sigma_U{}^i = d(du^i) + \sum du^j \wedge \theta_j{}^i$$

$$= 1/2 \sum (\Gamma^i_{jk} - \Gamma^i_{kj}) du^j \wedge du^k .$$

In this case if $D\sigma = 0$ the connection is called torsionless.

Example 2: If $\psi_V = \gamma_{UV}{}^{-1} \psi_U \gamma_{UV}$ is a tensor of type ρ = adjoint, then

$$D\psi_U = d\psi_U - \theta_U \psi_U + \psi_U \theta_U .$$

Thus far we have not considered whether connections exist. Note that if a connection has $\theta_U = 0$ then on $U \cap V$

$$\theta_V = -\gamma_{UV}^{-1} d\gamma_{UV} .$$

This suggests that we might be able to buy a very cheap connection by averaging this last description over a partition of unity, and in fact if $\{\lambda_U\}$ is a partition of unity subordinate to a cover and B_G is a principal G bundle defined by $\{\gamma_{UV}\}$ then

$$\theta_U(p) = - \sum_{W \ni p} \lambda_W \gamma_{WU}^{-1} d\gamma_{WU}$$

defines a connection.

Note if p lies in only one U of the covering then

$$\theta_U(p) = - \gamma_{UV}^{-1} d\gamma_{UV} = 0 .$$

In the case that the metric is induced by an immersion

$$x : M_n \longrightarrow \mathbb{R}^{m+p} ,$$

the conditions that a vector field Y be parallel with respect to the Levi-Civita connection has a simple geometric interpretation. In fact

$$x_*(DY) = dY|_{\text{Tangential}} .$$

In particular if two manifolds are tangent along a curve the derivative is independent of the choice of manifold containing the curve. As such we may envision the process of parallel translation of a vector field along a curve in the following way. Consider the submanifold generated by the tangent planes along the curve, this may be isometrically rolled out on euclidean space. In euclidean space the notion of parallel agrees with the euclidean notion hence in order to translate a vector field parallely, we roll out the tangential submanifold, move the vector field along the curve so that it remains parallel in the euclidean sense and then roll the tangential submanifold back along the original manifold.

The approach to connections thus far has been pragmatic and local. I want to mention a global interpretation which is important for the complex analytic category.

If G is a Lie group then its tangent bundle is again a Lie group. If

$$\phi : G \times G \longrightarrow G \quad \text{then} \quad \phi_* : T(G) \times T(G) \longrightarrow T(G)$$

and in fact

$$T(G) \simeq G x_{Ad} L(G) .$$

As such if $G \to B_G$ is a principal G-bundle then $T(G) \longrightarrow T(B)$ is a principal $T(G)$-bundle and considering $G \to T(G)$ as a subgroup via the zero section there is a short exact sequence of vector bundles over M which is due to Kobayashi and is called the Atiyah sequence given by

$$B_G \downarrow M \qquad\qquad T(B) \downarrow T(M)$$

$$0 \longrightarrow L(B) \longrightarrow T(B)/G \longrightarrow T(M) \longrightarrow 0 .$$

The space

$$T(B)/G \simeq T(B) x_{T(G)} T(G)/G \simeq T(B) x_{T(G)} L(G)$$

hence is locally given by strip maps

$$\phi_{T(U)} : T(U) \times L(G) \longrightarrow T(B)/G .$$

Matters being so a connection on B_G is a splitting of this sequence over M

$$T(B)\times_{T(G)}L(G) \underset{\Gamma}{\rightleftarrows} T(M) \longrightarrow 0$$

and the relation with the connection matrix defined before is given by

$$\Gamma(x(p)) = \phi_{T(V)}(x(p), -\theta_U(x(p))).$$

The unusual transformation law for the matrix θ_U arises since the action of $T(G)$ on $L(G)$ is non-linear.

The advantage of this approach besides the elegance is that there is an obstruction to splitting sequences of vector bundles which lies in

$$H^1(M, \underline{L(G) \otimes T^*(M)}) .$$

A detailed analysis of this cohomology set and the obstruction in the complex analytic case is carried out by Atiyah [1]. One of his main results is that the existence of a complex analytic connection implies the vanishing of all Chern classes if the manifold is compact and Kähler.

A second obvious question is whether

$$B_G{\times}_\rho A \xrightarrow{D} B_G{\times}_\rho A \otimes T^*(M) \xrightarrow{D} B_G{\times}_\rho A \otimes \Lambda^2 T^* M \xrightarrow{D} \cdots$$

defines a complex.

A local calculation shows

$$D^2\sigma|_U = \sigma_U \wedge \rho_*(\Theta_U) \quad \text{(Ricci Identity):}$$

where

$$\Theta_U = d\theta_U - \theta_U \wedge \theta_U .$$

The relation

$$D^2\sigma|_V = D^2\sigma|_U \rho(\gamma_{UV})$$

implies

$$\Theta_V = \gamma_{UV}^{-1}\Theta_U\gamma_{UV}$$

and since the elements of even degree form the center of an exterior algebra we may view the center as a commutative ring and form

$$\det(I + \lambda\Theta_U) = \sum P_i(\Theta_U)\lambda^i$$

but by the relation on $U \cap V$

$$P_i(\Theta_U) = P_i(\Theta_V) \quad \text{on} \quad U \cap V$$

hence there is a global form of degree $2i$ on M given by

$$P_i(\Theta)|_U = P_i(\Theta_U) \quad .$$

A fundamental property of these forms is that

$$dP_i(\Theta) = P_i(D\Theta)$$

and the classical Bianchi identity that

$$\begin{aligned} d\Theta &= d(d\theta - \theta \wedge \theta) \\ &= -d\theta \wedge \theta + \theta \wedge d\theta \\ &= -(d\theta - \theta \wedge \theta) \wedge \theta + \theta \wedge (d\theta - \theta \wedge \theta) \\ &= -\Theta \wedge \theta + \theta \wedge \Theta \end{aligned}$$

implies

$$D\Theta = d\Theta + \Theta \wedge \theta - \theta \wedge \Theta = 0$$

hence that $P_i(\Theta)$ are global closed $2i$ forms.

We define

$$\text{Pont } T(M) \subset H^*(M;\mathbb{R})$$

to be the subring generated by the de Rham classes of $P_i(\theta)$.

There is of course a topological definition of this subring but the differential geometric approach to the study of these classes is that as de Rham classes we have certain distinguished representatives which can be related to geometric data. As such various arguments are suggested which are not usually available to topologists.

Example 1: Connection deformation

$$\theta_t = (1 - t)\theta + t\overline{\theta} \qquad \theta, \overline{\theta} \text{ two connections}$$

then

$$\frac{d\, P_i(\theta_t)}{dt} = i\, d\, \tilde{P}_i(\overline{\theta} - \theta, \Theta_t, \ldots, \Theta_t)$$

where $\tilde{P}_i$ is the totally polarized multilinear form corresponding to the homogeneous form P_i.

In particular $\int_0^1 dt$ gives

$$P_i(\Theta) \sim P_i(\overline{\Theta})$$

and the classes are independent of the choice of connection.

Example 2: Choose θ = Levi-Civita connection in orthonormal frames then

$${}^t\theta = -\theta$$

which implies

$$P_i(\Theta) = 0 \quad i \text{ odd} .$$

The class

$$[P_{2i}(\Theta)] = i^{th} \text{ Pontryagin class} .$$

Example 3: Curvature Deformation. There is a matrix ψ of 1-forms of type adjoint on $B_{O(n)}$ such that

$$\Theta_t = \Theta + D(t\psi)$$

satisfies

$$\Theta_0 = \Theta \quad \Theta_1 = 0$$

and

$$\frac{d\, P_i(\Theta_t)}{dt} = i\ d\ P_i(\psi,\ \Theta_t,\ldots,\Theta_t)$$

and hence with $\int_0^1 dt$ we find

$$\pi^* P_i(\Theta) = d\ T\ P_i(\Theta)$$

where

$$\pi : B_{0(n)} \to M$$

and

$$TP_i(\Theta) = i \int_0^1 P_i(\psi,\ \Theta_t,\ldots,\Theta_t)dt\ .$$

$T\ P_i(\Theta)$ is called a transgression of the class $P_i(\Theta)$ from M to $B_{0(n)}$.

§3. Vanishing Theorems of Bott and Pasternak

We have seen in §0 that a codimension p foliation leads to coverings by distinguished neighborhoods such that on the overlap of two such neighborhoods U and V

$$dV = dU\gamma_{UV}$$

where

$$\gamma_{UV} = \left(\begin{array}{c|c} * & 0 \\ \hline * & g_{UV} \end{array}\right) .$$

Thus

$$dv^i = du^j (g_{UV})^i_j \qquad m + p - 1 \leq i,\ j \leq m$$

and as a result differentiation gives

$$0 = -\sum du^j \wedge d(g_{UV})^i_j$$

and hence $d(g_{UV})^i_j$ lies in the ideal I of forms which vanish on all of the leaves.

Now $\{g_{UV}\}$ are the transition functions for the subbundle E of the cotangent bundle $T^*(M)$ with trivialization over a distinguished neighborhood given by

$$E|_{U_\alpha} = \{du_\alpha^{m-p+1},\dots,du_\alpha^m\} .$$

The cheapest connection matrix for this bundle is given by

$$\theta_U = -\sum \lambda_W\, g_{WU}^{-1} dg_{WU}$$

all of whose entries lie in the ideal $I|_{U_\alpha}$. As such the associated curvature matrix

$$\begin{aligned}\Theta_U &= d\theta_U - \theta_U \wedge \theta_U \\ &= -\sum d\lambda_W g_{WU}^{-1} dg_{WU} - \sum \lambda_W dg_{WU}^{-1} \wedge dg_{WU} - \theta_U \wedge \theta_U\end{aligned}$$

also lies in the ideal $I|_{U_\alpha}$. Now (1) of the Lemma in §0 implies

$$\text{Ideal } I|_{U_\alpha} \cap \Lambda^i T^*(M) = 0 \quad \text{for} \quad i > 2p$$

and hence

$$P_i(\Theta) = 0 \quad \text{for} \quad i > 2p$$

and as a result we have Bott's theorem [2] that

$$\text{Pont}^i E = 0 \quad \text{for} \quad i > 2p .$$

Now if additional geometric information is available then it may be possible to choose a more sophisicated connection for which the curvature matrix

$$\Theta \in \text{subalgebra } (I|_{U_\alpha} \cap T^* M))$$

and hence for which

$$\text{Pont}^i E = 0 \quad \text{for} \quad i > p .$$

As an example let us consider the case of Riemannian foliations. We have seen that the transition functions for the frame bundle of a foliated manifold have the form

$$\gamma_{UV} = \left(\begin{array}{c|c} * & 0 \\ \hline * & g_{UV} \end{array}\right) .$$

If we choose any Riemannian metric then we may choose frames with legs either tangent to the foliation or normal to the foliation. This reduces the transition functions to the form

$$\gamma_{UV} = \left(\begin{array}{c|c} * & 0 \\ \hline 0 & g_{UV} \end{array}\right) .$$

If G is the matrix of the metric in such frames then

$$G = \left(\begin{array}{c|c} G' & 0 \\ \hline 0 & G'' \end{array}\right) .$$

If in addition the entries of G'' satisfy

$$dG'' \wedge du_\alpha^{m-p+1} \wedge \ldots \wedge du_\alpha^m = 0$$

on distinguished neighborhoods then the foliation is said to be a <u>Riemannian foliation</u>. Geometrically this means that the distance between two leaves is constant.

If we let

$$\sum G''_{ij}\, du_\alpha^i du_\alpha^j = \sum (\tau_\alpha^i)^2 \qquad (m + p - 1 \leq i, j \leq m)$$

then there is a unique matrix ϕ_j^k on U_α such that

$$d\tau^k = \sum \tau^i \wedge \phi_i^k$$

and

$$dG''_{ij} = \phi_i^k G_{kj} + G_{ik} \phi_j^k .$$

These matrices are easily seen to define a connection for E and to have a curvature matrix with every entry in the subalgebra generated by $I|_{U_\alpha} \cap T^*(M)$. This proves Pasternak's theorem [4] that a Riemannian foliation has

$$\mathrm{Pont}^i(E) = 0 \quad \text{for} \quad i > p.$$

References

1) M. Atiyah, "Complex Analytic Connections in Fibre Bundles", Trans. A.M.S. (1957) p. 181-207.

2) R. Bott, "On a Topological Obstruction to Integrability", Proc. Symp. Pure Math, Vol. 16, A.M.S. Providence, R.I. 1970 pp. 127-131.

3) C. Godbillon and J. Vey, "Un Invariant des Feuilletages de Codimension 1" C. R. Acad. Sci. Paris Ser. A-B273 (1971) pp. 92-95.

4) J. Pasternak, "Foliations and Compact Lie Group Actions" Comment. Math. Helv., 46 (1971) pp. 467-477.

5) J. Sondow, "The Godbillon-Vey Invariant of a Product Foliation is Zero", Dynamical Systems, Academic Press New York 1973 pp. 545-547.

6) B. Reinhart and J. Wood, "A Metric Formula for the Godbillon-Vey Invariant for Foliations", Proc. A.M.S. 38 (1973) pp. 427-429.

7) W. Thurston, "Noncobordant Foliations of S^3", Bull. A.M.S., 78 (1972) pp. 511-514.

8) W. Thurston and H. Rosenberg, "Some Remarks on Foliations" Dynamical Systems, Academic Press New York 1973, pp. 463-478.

PONTRYAGIN POLYNOMIAL RESIDUES OF ISOLATED FOLIATION SINGULARITIES

by Paul A. Schweitzer, S.J. and Andrew P. Whitman, S.J.*

Baum and Bott [1] have introduced and studied residues for singularities of holomorphic foliations. Each residue is associated with a polynomial in the Chern classes of the normal bundle to the foliation and measures the failure of the polynomial to satisfy Bott vanishing along the singular sets.

In this paper we define analogous Pontryagin polynomial residues for isolated singularities of real foliations. There is a similar theory for more general singularity sets which we hope to present elsewhere, but the case of isolated singularities is a little simpler (since the residues are real numbers, rather than homology classes) and includes the non-trivial examples which we know at present. The Residue Existence Theorem is stated and proved in §1 and the examples are presented in §2.

We would like to thank H. Shulman and A. Haefliger for helpful suggestions.

* Both authors gratefully acknowledge the support of FINEP and CNPq (Brazil) during this research.

1. The Residue Existence Theorem

Let M be a smooth oriented manifold of dimension $4k$ and S a finite subset. By a <u>singular foliation</u> F of M with <u>singular set</u> S we mean a smooth foliation of $M-S$. Let Q be the normal bundle to F over $M-S$ and let $p_j(Q) \in H^{4j}(M-S;\mathbb{R})$ be its jth real Pontryagin class. The cohomology class $\bar{p}_j = (i^*)^{-1}p_j(Q) \in H^{4j}(M;R)$ is well defined for $j=1,2,\ldots,k-1$ since the inclusion $i: M-S \longrightarrow M$ induces cohomology isomorphisms below dimension $4k-1$. Let $\psi \in \mathbb{R}[X_1,\ldots,X_{k-1}]$ be a weighted homogeneous polynomial of weighted degree $4k$ where $\deg X_j = 4j$.

Assume from now on that the codimension of F is $q < 2k$. If $S = \emptyset$ then the Bott vanishing theorem states that $\psi(p_1(Q),\ldots,p_{k-1}(Q)) = 0$. If we allow S to be non-empty we have:

THE RESIDUE EXISTENCE THEOREM (See [1], Theorem 2). For each point $x \in S$ there is defined a residue $\mathrm{Res}_\psi(F,x) \in \mathbb{R}$ with the following properties:

(1.1) The residue is <u>local</u>, i.e. if U is an open set of M containing x, then $\mathrm{Res}_\psi(F|U,x) = \mathrm{Res}_\psi(F,x)$.

(1.2) The residue is a diffeomorphism invariant, i.e. if $f: U' \longrightarrow U$ is a diffeomorphism taking x' to x and F is a foliation of $U-\{x\}$ inducing $f^{-1}F$ on $U'-\{x'\}$, then $\mathrm{Res}_\psi(f^{-1}F,x') = \pm\mathrm{Res}_\psi(F,x)$, with the sign depending on whether f preserves or reverses the orientation.

(1.3) If M is compact, then $\sum_{x \in S}\mathrm{Res}_\psi(F,x) = \langle \psi(\bar{p}_1,\ldots,\bar{p}_{k-1}),[M]\rangle$, where $\langle \cdot,[M]\rangle$ denotes the Kronecker product with the orientation homology class of M.

<u>Remarks</u>. (1.4) It is not difficult to extend (1.2) to invariance under concordance (in particular, therefore, under integrable homotopy) and, if $q < 2k-1$, under homotopy of foliations (cf. Heitsch rigidity [5]), provided the singular set remains fixed.

(1.5) If M is compact and F has only one singular point x, then $\mathrm{Res}_\psi(F,x)$ must be an integer, by (1.3). Can the residue for a given polynomial ψ and codimension q take on all real values? It is reasonable to conjecture that in the rigid range ($q < 2k-1$) the residue will be rational [1, Rationality conjecture, p. 287].

Before proving the Theorem we recall some formal properties of the Chern construction of the real Pontryagin classes and basic connections. A connection ∇ on a smooth q-plane bundle Q over U determines a closed differential $4j$-form $\alpha_j = \alpha_j(\nabla) \in \Lambda^{4j}(U)$ whose de Rham cohomology class is $[\alpha_j] = p_j(Q)$ [2, p. 29]. The definition is local in that $\alpha_j(\nabla)|V = \alpha_j(\nabla|V)$ if V is open in U. If Q is the normal bundle to a foliation and ∇ a basic connection then (always assuming weighted deg $\psi > 2q$) we have Bott vanishing:

(1.6) $\quad \psi(\alpha_1,\ldots,\alpha_{k-1}) = 0 \in \Lambda^{4k}(U) \quad$ [2, pp. 33-35].

We remark also that basic connections are extendable, i.e. a basic connection defined on a neighborhood of a closed set, if restricted to a smaller neighborhood, will extend to a basic connection on all of U. (Just use convexity of the space of basic connections and a partition of unity argument).

Definition of the Residue. We shall define $\mathrm{Res}_\psi(F,x)$ for $x \in S$. We assume the hypotheses leading up to the Theorem. By replacing M by a smaller neighborhood of x we may assume that $S = \{x\}$. Now choose

(i) a basic connection ∇ on the normal bundle Q to F over $M-S$, and

(ii) an open neighborhood W of x whose closure $\overline{W}$ is diffeomorphic to the closed unit disk in $\mathbb{R}^{4k}$.

Extend the form $\alpha_j = \alpha_j(\nabla)|\, M-W$ to a closed form $\overline{\alpha}_j \in \Lambda^{4j}(M)$, $j = 1,\ldots,k-1$. This is possible by a simple cochain argument, as follows. Extend α_j to any form $\beta_j \in \Lambda^{4j}(M)$. Then $d\beta_j$ is closed and vanishes outside W. By the Poincaré lemma (or, equivalently, since $H^{4j+1}_{dR}(M,M-W) = 0$), there exists a form $\gamma_j \in \Lambda^{4j}(M)$ vanishing outside W such that $d\gamma_j = d\beta_j$. Then $\overline{\alpha}_j = \beta_j - \gamma_j$ is the desired extension.

We now define the residue at x to be

(1.7) $\quad \mathrm{Res}_\psi(F,x) = \int_M \psi(\overline{\alpha}_1,\ldots,\overline{\alpha}_{k-1}) \in \mathbb{R}.$

The integral is defined since by (1.6) the form $\psi(\overline{\alpha}_1,\ldots,\overline{\alpha}_{k-1})$ vanishes outside $\overline{W}$.

We must show that the residue is independent of the choices of the $\bar{\alpha}_j$'s, W, and ∇.

For a fixed index j, let $\bar{\alpha}_j'$ be another closed extension of α_j. Then $\bar{\alpha}_j - \bar{\alpha}_j' = d\gamma$ for some form $\gamma \in \Lambda^{4j-1}(M)$ supported on $\bar{W}$ (since $H^{4j}_{dR}(M,M-W) = 0$). Furthermore there is a factorization

$$\psi(\bar{\alpha}_1,\ldots\bar{\alpha}_j,\ldots,\bar{\alpha}_{k-1}) - \psi(\bar{\alpha}_1,\ldots\bar{\alpha}_j',\ldots,\bar{\alpha}_{k-1}) = (\bar{\alpha}_j-\bar{\alpha}_j')\omega$$

where ω is a closed form (in fact a polynomial in the $\bar{\alpha}_\ell$'s and $\bar{\alpha}_j'$), so

$$\int_M(\bar{\alpha}_j - \bar{\alpha}_j')\omega = \int_{\bar{W}} d(\gamma\omega) = \int_{\partial\bar{W}}\gamma\omega = 0$$

since γ vanishes on $\partial\bar{W}$. Thus replacing $\bar{\alpha}_j$ by $\bar{\alpha}_j'$ leaves the residue unchanged.

(We are grateful to William Thurston for suggesting the essential idea of this "linking number" argument.)

Now hold ∇ fixed and let W vary. If $W_1 \subset W_2$, we may use the form $\bar{\alpha}_j$ extending $\alpha_j(\nabla)|M-W_1$ as the extension of $\alpha_j(\nabla)|M-W_2$ as well. Thus the residue does not depend on W.

Finally let ∇_1 and ∇_2 be two basic connections on Q. Choose $\bar{W}_1 \subset W_2$. Then by extendability of basic connections we can manufacture a new basic connection ∇ extending both $\nabla_2|M-W_2$ and $\nabla_1|\bar{U}-\{x\}$ where U is a neighborhood of $\bar{W}_1$ with $\bar{U} \subset W_2$. Then

$$\mathrm{Res}_{\nabla_1, W_1} = \mathrm{Res}_{\nabla, W_1} = \mathrm{Res}_{\nabla,W_2} = \mathrm{Res}_{\nabla_2,W_2}$$

where of course $\mathrm{Res}_{\nabla,W}$ denotes the value of the residue defined using ∇ and W.

<u>Proof of the Theorem</u>. From the preceding discussion it is clear that the residue is well-defined and local (1.1). To see (1.2), it suffices to note that at every step the construction pulls back to U' under the diffeomorphism $f: U' \longrightarrow U$, and that changing the orientation of M merely changes the sign of the integral (1.7).

For (1.3) let M be compact, let ∇ be a basic connection on Q over M-S, and for each $x \in S$ choose a disk neighborhood W_x of x such that the W_x's have pairwise disjoint closures. Let $\bar{\alpha}_j \in \Lambda^{4j}(M)$ be a closed form extending $\alpha_j(\nabla)|M_o$, where $M_o = M - \bigcup_{x\in S} W_x$. According to the Chern construction, $\alpha_j(\nabla)|M_o$ represents the Pontryagin class $p_j(Q|M_o) \in H^{4j}(M_o;\mathbb{R})$. The inclusions $M_o \xrightarrow{i_o} M\text{-}S \xrightarrow{i} M$ induce isomorphisms on cohomology in dimension $4j$, so from

$$i_o^* i^* [\bar{\alpha}_j] = [\bar{\alpha}_j(\nabla)|M_o] = p_j(Q|M_o) = i_o^* p_j(Q)$$

we conclude that $[\bar{\alpha}_j] = (i^*)^{-1} p_j(Q) = \bar{p}_j \in H^{4j}(M;\mathbb{R})$. Finally

$$\begin{aligned}
\sum_{x\in S} \mathrm{Res}_\psi(F,x) &= \sum \int_{\bar{W}_x} \psi(\bar{\alpha}_1,\ldots,\bar{\alpha}_{k-1}) \\
&= \int_M \psi(\bar{\alpha}_1,\ldots,\bar{\alpha}_{k-1}) \\
&= \left\langle [\psi(\bar{\alpha}_1,\ldots,\bar{\alpha}_{k-1})], [M] \right\rangle \\
&= \left\langle \psi(\bar{p}_1,\ldots,\bar{p}_{k-1}), [M] \right\rangle.
\end{aligned}$$

This completes the proof of the Theorem.

Remarks (1.8). If F extends to a foliation of $M\text{-}S\cup\{x\}$, then $\mathrm{Res}_\psi(F,x) = 0$, for in this case we may choose a basic connection ∇_x over $M\text{-}S\cup\{x\}$ and let $\bar{\alpha}_j = \alpha_j(\nabla_x)$ so that $\psi(\bar{\alpha}_1,\ldots,\bar{\alpha}_{k-1})$ will vanish on $\bar{W}_x$.

(1.9) There is a corresponding residue theory for isolated singularities of a Γ_q structure γ on $M\text{-}S$. The normal bundle Q is replaced by the normal bundle to the graph $gr(\gamma)$ of γ [3, p. 188]. Let $\tilde{\nabla}$ be a basic connection on $\tilde{Q}$ for the canonical foliation of $gr(\gamma)$, and replace $\alpha_j(\nabla)$ by $s^*\alpha_j(\tilde{\nabla}) \in \Lambda^{4j}(M\text{-}S)$, where s is the zero section of $gr(\gamma)$ (or a smooth approximation thereof). The rest of the theory then proceeds as before. If γ defines a smooth foliation F_γ, the residue theories for foliations and for Γ_q structures yield the same residue $\mathrm{Res}_\psi(\gamma,x) = \mathrm{Res}_\psi(F_\gamma,x)$.

2. Examples for the Polynomials p_ℓ^2

Let $M = S^{4\ell} \times S^{4\ell}$ (so that $k = 2\ell$) and fix $x_0 \in M$.

Proposition 2.1 If $q = 4\ell - 2 > 2$ or $q = 4\ell - 1$, then there is a codimension q foliation F of $M_0 = S^{4\ell} \times S^{4\ell} - \{x_0\}$ such that

$$\mathrm{Res}_\psi(F, x_0) \neq 0$$

for the polynomial $\psi(p_1, \ldots, p_{k-1}) = p_\ell^2$.

Corollary 2.2 If $\eta \in \pi_{4\ell}(B\Gamma_q)$ detects p_ℓ and $q = 4\ell - 2$ or $4\ell - 1$, then the Whitehead product $[\eta, \eta] \in \pi_{8\ell-1}(B\Gamma_q)$ has infinite order.

Haefliger has also proven (2.2) in [4,§5]. Before beginning the proofs we make the following remark.

Remark (2.3) Proposition 2.1 remains true if $S^{4\ell} \times S^{4\ell}$ is replaced by a compact oriented 8ℓ dimensional manifold M such that

(i) $H^{4\ell}(M;\mathbb{R}) \neq 0$

(ii) M is $8\ell-q-3$ connected, where $q = 4\ell-2 > 2$ or $q = 4\ell-1$.

The proof is essentially the same. Two examples are the quaternionic and Cayley projective planes.

Proof of Proposition 2.1 The proof, in four steps, is an existence proof which unfortunately gives little geometric insight into the foliation F.

1. There is a q-plane bundle Q over M_0 such that $\bar{p}_\ell^{\,2} \neq 0 \in H^{8\ell}(S^{4\ell} \times S^{4\ell}; \mathbb{R})$.

To see this, note that M_0 homotopy retracts onto its 4ℓ-skeleton $S^{4\ell} \vee S^{4\ell}$ (chosen such that $x_0 \notin S^{4\ell} \vee S^{4\ell}$). We claim that there exists a map $f\colon S^{4\ell} \longrightarrow BO_q$ detecting p_ℓ, i.e. such that $f^*p_\ell \neq 0 \in H^{4\ell}(S^{4\ell}; \mathbb{R})$. Then it will suffice to let $f_Q\colon M_0 \longrightarrow BO_q$ be the composition of the retraction $M_0 \longrightarrow S^{4\ell} \vee S^{4\ell}$ with $f \vee f\colon S^{4\ell} \vee S^{4\ell} \longrightarrow BO_q$, and let Q be the bundle with classifying map f_Q. The class $\bar{p}_\ell^2 \neq 0$ since $\bar{p}_\ell$ is non-vanishing on both factors $S^{4\ell}$.

To construct f, note that the Stiefel manifold $V_{2j+1,2} = O_{2j+1}/O_{2j-1}$ is a rational $4j-1$ sphere. By induction on j in the exact rational homotopy sequence of the fibration $V_{2j+1,2} \longrightarrow BO_{2j-1} \longrightarrow BO_{2j+1}$

we find that the rational homotopy groups $\pi_*(BO_{2\ell+1}) \otimes Q$ have one generator in each dimension $4j$, $j=1,\ldots,\ell$. A generator of $H^{4\ell-1}(V_{2\ell+1,2};\ \mathbb{R})$ transgresses to a non-zero multiple of $p_\ell \in H^{4\ell}(BO_{2\ell+1};\ \mathbb{R})$. Consequently a generator of $\pi_{4\ell}(BO_{2\ell+1}) \otimes Q$ detects p_ℓ. An appropriate multiple of this generator comes from a map $S^{4\ell} \longrightarrow BO_{2\ell+1}$ which, composed with the inclusion $BO_{2\ell+1} \longrightarrow BO_q$ (note that $2\ell+1 \leq q$), yields the desired f.

2. The q-plane bundle Q can be embedded in the tangent bundle T of M_o. We apply an obstruction theory of E. Thomas [7, p.657 or 6, pp. 344-347] to show that there exists a splitting $T \cong E \oplus Q$. Consider the map $\pi = (\mu, p_2) : BO_{8\ell-q} \times BO_q \longrightarrow BO_{8\ell} \times BO_q$ where $\mu: BO_{8\ell-q} \times BO_q \longrightarrow BO_{8\ell}$ is the Whitney product map and p_2 is the projection on the second factor. The homotopy theoretic fiber of the map π is $V_{8\ell,q} = O_{8\ell}/O_{8\ell-q}$ which is $8\ell-q-1$ connected. Since $8\ell-q-1 \geq 4\ell$ and M_o has the homotopy type of its 4ℓ skeleton, there is a homotopy lifting

$$\begin{array}{ccc} & & BO_{8\ell-q} \times BO_q \\ & \overset{(f_E,f_Q)}{\nearrow} & \downarrow \\ M_o & \xrightarrow{(f_T,f_Q)} & BO_{8\ell} \times BO_q \end{array}$$

of (f_T,f_Q) to (f_E,f_Q), where f_V denotes the classifying map of the bundle V, $V = T$, E, or Q. Then $\mu\circ(f_E,f_Q) \simeq f_T$ so $T \cong E \oplus Q$.

3. There is a Γ_q structure γ on M_o with normal bundle isomorphic to Q. The homotopy theoretic fiber $B\overline{\Gamma}_q$ of $B\Gamma_q \longrightarrow BO_q$, the classifying map of the universal Γ_q structure, is $q+1$ connected ([8], Theorem 2; cf. [3], 3.1). Hence $f_Q: M_o \longrightarrow BO_q$ lifts to $f_\gamma: M_o \longrightarrow B\Gamma_q$.

4. There is a foliation F of M_o with normal bundle isomorphic to Q and consequently $\mathrm{Res}_\psi(F,x_o) \neq 0$. The map $(f_\gamma,f_E): M_o \longrightarrow B\Gamma_q \times BO_{8\ell-q}$ is a homotopy lifting of $f_T: M_o \longrightarrow BO_{8\ell}$ for the composition $B\Gamma_q \times BO_{8\ell-q} \longrightarrow BO_q \times BO_{8\ell-q} \xrightarrow{\mu} BO_{8\ell}$. According to Theorem 2 of [3], on an open manifold the homotopy classes of such liftings are in one-to-one correspondence with integral homotopy classes of foliations. Therefore γ is homotopic to a foliation F of M_o, and the normal bundle to F is isomorphic to Q. Finally by (1.3) we have

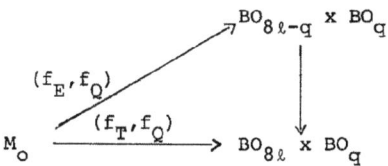

$$\mathrm{Res}_{\psi}(F,x_o) = \langle \bar{p}_{\ell}^2, [M] \rangle \neq 0,$$

completing the proof.

<u>Proof of Corollary 2.2</u> We use the corresponding residue theory for Γ_q structures (1.9). Let $f: S^{4\ell} \longrightarrow B\Gamma_q$ be a representative of η and let f_γ be the composition $M_o \xrightarrow{\text{retraction}} S^{4\ell} \vee S^{4\ell} \xrightarrow{\text{fvf}} B\Gamma_q$. The Γ_q structure γ classified by f_γ has a singularity at x_o and as in the preceding argument $\mathrm{Res}_{\psi}(\gamma, x_o) \neq 0$. If the restricted map $f_\gamma | \partial \bar{W}$ were null homotopic (where W is a disk neighborhood of x_o), then $f_\gamma | M-W$ would extend to a map $\bar{f}_\gamma: M \longrightarrow B\Gamma_q$ and the residue would vanish as in (1.8). Therefore the homotopy class $[\eta,\eta] \in \pi_{8\ell-1}(B\Gamma_q)$ of $f_\gamma | \partial\bar{W}$ cannot vanish. The same argument applied to $r\eta$ for $r > 0$ shows that $r^2[\eta,\eta] = [r\eta, r\eta] \neq 0$ so that $[\eta,\eta]$ must have infinite order.

Pontificia Universidade Catolica do Rio de Janeiro.

REFERENCES

[1] P. Baum and R. Bott, Singularities of holomorphic foliations, J. Differential Geometry, 7 (1972), 279-342.

[2] R. Bott, Lectures on characteristic classes and foliations (notes by L. Conlon), Springer Lecture Notes in Math. 279 (1972) 1-94.

[3] A. Haefliger, Feuilletages sur les variétés ouvertes, Topology 9, (1970), 183-194.

[4] ___________, Whitehead products and differential forms, these Proceedings.

[5] J. Heitsch, Deformations of secondary characteristic classes, Topology 12 (1973), 381-388.

[6] E. Thomas, Fields of tangent k-planes on manifolds, Invent. Math. 3, (1967), 334-347.

[7] ________, Vector fields on manifolds, Bull. Amer. Math. Soc. 75, (1969), 643-683.

[8] W. Thurston, Foliations and groups of diffeomorphisms, Bull. Amer. Math. Soc., 80 (1974), 304-307.

STRUCTURES FEUILLETEES

Pourquoi les a-t-on étudiées. Comment les a-t-on étudiées. Est-il "rentable" de continuer ces investigations ?

par

G. REEB

L'exposé débute inévitablement par l'enquête : Y a-t-il un botaniste dans l'assistance ? En effet, depuis toujours, (M. Maresquelle a initié la série en 1948) un botaniste au moins se trouve interpelé par les vocables : feuilles, feuillage,... La même assimilation a permis à un typographe, exaspéré, de composer malicieusement un titre : feuilles mortes.

A défaut de botaniste l'assemblée comptera peut-être un pâtissier. La pâte feuilletée - j'ai de bonnes raisons de le croire - donne une bonne idée d'un feuilletage (de codimension 1 dans $\mathbb{R}^3$) dont elle dessine bien les feuilles et en suggère des propriétés.

1. ORIGINES.

Tout a commencé (vers 1935) par une question - inédite à ma connaissance - de H. HOPF :

Constatant que sur la sphère "euclidienne" S_3 (et plus généralement sur les variétés orientables de dimension 3) il existe des systèmes de p champs de vecteurs linéairement indépendants en tout point [en abrégé : p-champ] pour $p = 1, 2, 3$, il apparaît opportun de poser le problème suivant :

$\underline{Q_1}$. Existe-t-il sur S_3 (ou sur V_3) un 1-champ complètement intégrable $\vec{V}$? [i.e. tel que $\vec{V}\,\vec{\mathrm{rot}}\,\vec{V} \equiv 0$ (notations de Hopf, voir plus bas Q_1')] ?

H. Hopf était, à l'évidence, animé du souci suivant : trouver un principe de classification raisonnable, ou tout au moins non trivial, - au-delà des invariants d'homologie ou d'homotopie alors connus pour les variétés compactes à trois dimensions. (Ces variétés échappaient alors, et échappent encore, à des tentatives - pourtant séduisantes - de classification par des critères tels que : existence de p - champs,...). On sait que ce critère échoue également : les variétés compactes de dimension 3 peuvent être "feuilletées"? Faut-il éprouver comme une déception le fait (relativement récent) que toutes les sphères S_{2n+1} admettent une structure feuilletée de co-dimension un ?

Notons ici la question [connue comme "le problème du rang"] :

$\underline{\underline{Q_2}}$. Existe-t-il sur S_3 (ou V_3) un 2 - champ correspondant aux transformations infinitésimales (de base) du groupe de Lie R^2 opérant sur S_3 ?

Cette question dont les généralisations sont maintenant bien familières aux géomètres, cette question donc n'a pas été formulée - à ma connaissance du moins - par H. Hopf, ni par les géomètres de son temps.

Toujours est-il que Q_1 a aussitôt retenu l'attention de Ch. EHRESMANN qui a mis ce problème "en conserve" dans un magnifique registre-répertoire que mon maître m'avait autorisé à consulter en 1942, non sans apporter une certaine solennité - justifiée - à la communication du registre. Comment taire ici un regret : le riche contenu du répertoire ne semble pas avoir été porté à la connaissance d'un public élargi ? On sait comment Ch. Ehresmann s'est piqué au jeu et a apporté depuis 1942 de puissantes impulsions à Q_1 .

Ce rappel, concernant l'origine de l'étude des structures feuilletées [qu'on me permette ici de revendiquer le choix du vocable souligné, vocable qui connut une heureuse fortune (jusqu'à l'honneur d'être cité par ETIEMBLE* en compagnie d'autres termes comme fibré,... en modèle) ; en échange de cette permission je promets de ne pas parler de G.R. par la suite], ce rappel donc, permet de bien circonscrire notre sujet :

<u>Lorsque nous parlons de structure feuilletée ici, nous aurons présent à l'esprit le cas où la dimension des feuilles est au moins 2 (i.e. le cas où des conditions de complète intégrabilité - Théorème de Frobenius si on préfère - interviennent effectivement)</u>.

Bien sûr le cas où la dimension des feuilles est 1 (i.e. le cas des trajectoires d'un 1-champ) a de brillantes lettres de noblesse, il a été illustré par des noms célèbres et toute une HISTOIRE qu'il n'y a pas lieu de rappeler ici. De plus le langage propre aux feuilletages n'a pas manqué d'apporter quelque éclairage nouveau, même dans ce cas. Mais en un mot, comme en mille, l'introduction des conditions de complète intégrabilité produit des phénomènes entièrement nouveaux - et c'était là l'intuition brillante de H. Hopf.

Il y a lieu de remarquer ici, qu'à diverses reprises, des chercheurs ont réinventé à leur tour, vers les années 55 à 60 l'étude des structures feuilletées. Ces tentatives (non publiées) étaient souvent axées sur l'idée excellente, mais prématurée, de retrouver un théorème de "Poincaré Bendixon". Ces essais pourraient figurer dans la liste $P_1 \dots P_7$; ils aboutissent logiquement à l'étude des "bouts" des feuilles.

Parlant <u>d'intuition</u>, il faut saisir ici l'occasion d'opposer d'une part l'intuition sûre (prospective en tout cas) de géomètres tels que Hopf, Ehresmann, de Rham, Bouligand,...(pour en rester à notre sujet) et d'autre part l'intuition défaillante (et par conséquent paralysante) d'autres géomètres

---

* "Le jargon des sciences".

- dont les uns n'ont eu que le temps de donner un coup d'oeil superficiel au sujet.

Il n'est pas inutile de préciser ceci par quelques anecdotes :

Reformulons Q_1 sous la forme plus adéquate et duale Q'_1 .

Q'_1 Existe-t-il sur S_3 (ou S_{2n+1} ou V_n) une forme de Pfaff ω vérifiant les conditions suivantes $\omega \wedge d\omega \equiv 0$ et $\omega_x \neq 0$ en tout point x ?

et énonçons un nouveau problème :

Q_3 : Etudier les intégrales (feuilles) de l'équation $\omega = 0$ où ω répond au problème Q'_1 .

Consultés sur les problèmes Q'_1 ou Q_3 la plupart des géomètres ont eu pendant longtemps (disons jusqu'en 1965) une réaction assez semblable à celle-ci."La question Q_2 doit pouvoir se ramener plus ou moins naturellement (entendez par des méthodes inspirées des théorèmes de de Rham) au problème suivant :

Q_4 : Existe-t-il une forme ω fermée, avec $\omega_x \neq 0$ en tout point x ?

et Q_3 est dès lors banal". (En effet argumentera-t-on, ω admet localement un facteur intégrant ; il n'y a plus qu'à recoller). Cette affaire est en fait liée à la découverte de l'invariant de Godbillon-Vey. Notons enfin que même Q_4 donne lieu à des développements intéressants (Tischler).

En fait, cette intuition erronée correspond tout de même à une structure célèbre : $d\omega = \bar{\omega} \wedge \omega$, $d\bar{\omega} = 0$. (cf. M_3). Il est rentable de démystifier de telles erreurs, ne serait-ce que pour apprendre à un jeune chercheur à maintenir, contre vent et marée, la direction qu'il aura choisie avec un "flair" sûr.

Voici un autre exemple - éloigné en apparence seulement - de notre sujet - où en raison de préjugés hâtifs mais généralisés, des développements ont été longtemps arrêtés ou du moins retardés.

Rappelons une définition : ω étant une forme de Pfaff, on appelle <u>classe</u> de ω au point x l'entier p_x pour lequel le premier terme de rang p_x dans la suite :

$$d\omega_x\ ,\ \omega_x \wedge d\omega_x\ ,\ [d\omega_x]^2\ ,\ \omega_x \wedge [d\omega_x]^2\ ,\ \dots$$

est nul. [On suppose toujours $\omega_x \neq 0$] .

"Complète intégrabilité" est donc équivalent à $p_x \leq 2$; cette notion est donc un maillon d'une chaîne qui partant de la notion de forme fermée aboutit au cas général.

La classe joue pour les formes un rôle analogue au rang pour les applications. On sait que cette analogie a été largement exploitée (Calabi, Martinet,...) et qu'un vaste champ de recherche est encore ouvert.

Cependant, une autre analogie : celle qui juxtapose 1 - champ et forme de Pfaff a permis à l'idée - fausse - suivante de se perpétuer :

"Localement les 1-champs sont, pour une dimension donnée, tous isomorphes ; il doit bien en être plus ou moins de même pour les formes" (ceci malgré E. Cartan et malgré la connaissance de l'oeuvre de Cartan par les colporteurs de la fausse nouvelle !).

Notons également que par un surprenant "retour" les trouvailles de Varela-Lutz redonnent quelque peu raison à cette opinion : toute forme de Pfaff ω est C_0-proche d'une expression $\lambda df + dg$ (globalement).

Un troisième exemple est donné par l'affirmation souvent entendue au début de la théorie : Une structure feuilletée est une structure par trop pauvre. Le cours de Haefliger, ici même, donne une conclusion définitive.

2. MOTIVATIONS.

H. Hopf avait vraisemblablement des motivations profondes (j'entends des motivations externes au sujet proprement dit, ou encore des motivations permettant d'espérer des applications de la théorie à naître) pour proposer l'étude des structures feuilletées ; mais qui saurait ou pourrait rapporter sur ce point ? Toujours est-il qu'une des meilleures propagandes pour une théorie réside précisément dans la "motivation". Or il faut bien le dire, il y a trente ans on ne voyait pas bien clairement la "motivation" de l'étude des feuilletages. Curieusement la situation s'est retournée, et sans trop de peine, on peut amorcer une liste de "motivations" encore qu'il me semble que les mathématiciens n'attachent pas assez de prix à ces arguments.

M_1 : Etude des systèmes différentiels ordinaires, holomorphes dans le champ complexe. On sait l'importance de l'oeuvre de P. PAINLEVE ; mais peu de gens se doutent que Painlevé a vraiment senti qu'un langage ad-hoc (celui des feuilletages) est pratiquement indispensable pour discourir en géomètre sur ce sujet. Cette motivation entraîne dans le présent un considérable mouvement de recherche (voir les investigations récentes de Moussu, Malgrange, sur le théorème de Frobenius complexe).

M_2 : Voici encore plus curieux : de nombreux résultats fort précis (concernant les minimaux exceptionnels par exemple) sont séculaires ! Raymond au hasard de ses lectures de POINCARE, RIEMANN, SCHOTTKY, s'est aperçu que la théorie des groupes fuchsiens permet immédiatement des constructions de feuilletages riches : en particulier on peut aboutir ainsi - non sans quelque chirurgie difficile - à un feuilletage de co-dimension un de S_3 admettant un minimal exceptionnel.

$\underline{\underline{M_3}}$: La théorie des actions de groupes de Lie (théorie bien plus ancienne que celle des feuilletages) conduit souvent à considérer des feuilletages engendrés. De même la théorie du "repère mobile" (CARTAN) ("duale" en un sens assez vague de la précédente) suggère des classes de feuilletages à structure transversale remarquable.

$\underline{\underline{M_4}}$: La thermodynamique a habitué de longue date la physique mathématique [cf. DUHEM P.] à la considération de formes de Pfaff complètement intégrables : la chaleur élémentaire dQ [notation des thermodynamiciens] représentant la chaleur élémentaire cédée dans une modification infinitésimale réversible est une telle forme complètement intégrable. Ce point ne semble guère avoir été creusé depuis lors.

$\underline{\underline{M_5}}$: La théorie des flots d'Anosov, conduit de la manière la plus naturelle à des structures feuilletées.

$\underline{\underline{M_6}}$: La géométrie "intégrale" aboutit également à quelques problèmes de feuilletages.

$\underline{\underline{M_7}}$: Last but not least, il semble que la géométrie algébrique moderne, dans l'investigation des variétés complexes compactes, mais non algébriques s'oriente vers l'étude des feuilletages se substituant aux classiques "cycles". [Ramis, Norguet, ont attiré l'attention sur ce point]. Affaire à suivre ?

$\underline{\underline{M_8}}$: On sait l'utilité en divers points de géométrie différentielle de théorèmes qui disent d'une variété compacte sur laquelle on a une fonction numérique, n'admettant que deux points critiques, qu'elle est une sphère. Il est légitime d'annexer ces résultats aux feuilletages. Il est encore plus légitime d'attribuer ce résultat à M. MORSE (inédit) et à ELSGOLTZ (1940 !).

$\underline{\underline{M_9}}$: Les diverses formes, même très sophistiquées, des théorèmes de stabilité doivent bien conduire quelque jour à des applications.

Au vu de cette liste - non exhaustive, pour sûr - une prospective, ne comportant guère de risque, semble se dégager : après le développement quelque peu explosif qu'a connu et que connaît l'étude des structures feuilletées pour elles-mêmes on verra probablement un développement tout aussi important des "motivations" et des "applications".

3. <u>EST-IL "RENTABLE" DE CONTINUER CES INVESTIGATIONS</u> ?

A en juger par les développements présents (il est facile de dénombrer une centaine d'auteurs productifs) la réponse à la question semble être "oui". Par contre il est vain de prétendre énumérer, ou simplement classer, les tendances de recherche actuelle, et, encore davantage, de tenter dégager les essais qui iront en s'affermissant. Mais il peut être possible de discerner <u>quelques types de problèmes, n'ayant guère retenu l'attention</u> des chercheurs, et dont il est raisonnable d'attendre un bon rendement.

Avant de donner quelques exemples, une remarque liminaire s'impose : Il est frappant de voir - à côtés de développements nécessaires mais techniquement difficiles - de nombreux résultats fondamentaux procéder d'idées très simples, voire naîves. Voici, pour mémoire, quelques exemples suggérés par de brèves indications : Tischler sur les formes fermées, l'invariant de Godbillon-Vey lié somme toute à la notion ancienne de "dernier multiplicateur", les exemples de Lutz-Varela, Raymond, Hector, dont il a été question plus haut, etc... ; ici il s'agit d'idées simples, démonstration incluse ; des idées qui auraient dû sauter aux yeux des premiers chercheurs.

D'autres idées, par exemple celle qui est à la base de la thèse de Haefliger sont simples, mais leur mise en oeuvre peut s'avérer difficile. Ici encore la liste serait longue : Bott, obstructions à l'intégrabilité, Sullivan, Thurston,... Il y a fort à parier que d'autres trouvailles aussi simples se révèleront payantes.

Voici la petite liste promise plus haut :

$\underline{\underline{P_1}}$: Les travaux évoqués en M_1 , M_7 touchent à un domaine dont la prospérité semble garantie pour longtemps (mais ceci c'est presque de la prospective à posteriori et partant trop facile). La suggestion suivante corrigera quelque peu cette dernière impression ; l'étude des fonctions abéliennes du point de vue des beaux feuilletages qui leur sont associés pourrait ouvrir l'accès à des recherches nouvelles. C'est là que le point de vue de P. Painlevé sur les fonctions abéliennes est le plus proche du point de vue auquel nous faisons allusion. D'autre part l'occasion est bonne d'insister sur ceci : Painlevé était vraiment le premier à recommander l'étude globale des systèmes de Pfaff complètement intégrables.

$\underline{\underline{P_2}}$: Songeant aux nombreux travaux sur les bouts des feuilles, les phénomènes à la DENJOY, SACKSTEDER , une notion "d'homologie de petits cycles" semble se dégager. Pourquoi ne pas recourir aux méthodes proposées par ROBINSON ?

$\underline{\underline{P_3}}$: La théorie du "contrôle optimal" suggère des échanges fructueux avec la théorie des feuilletages.

$\underline{\underline{P_4}}$: L'étude de la "classe" des formes ; $\underline{\underline{P_5}}$ la codimension ≥ 2 .
$\underline{\underline{P_6}}$.. Les aspects relevant de la topologie générale...

$\underline{\underline{P_7}}$: Le fameux problème de Hilbert, dont l'étude a été abordée par Petrowski et Landis. (Il s'agit de majorer le nombre de cycles limites de : $P\,dx + Q\,dy = 0$ où P, Q sont des polynômes en x et y) . Ce problème est toujours payant.

Notre propos était de parler des aspects d'une théorie qui précisément "ne font pas assez parler d'eux". Il n'est donc pas de notre devoir de donner une bibliographie, ni de citer systématiquement les contributions importantes.

Rigidity of the Centralizers of Diffeomorphisms and Structural Stability of Suspended Foliations

J. Palis

In this paper we relate some known and some new results about the centralizer group of a diffeomorphism. These results have direct application on the structural stability of suspended foliations and throw a bit of light into the formidable study of $\mathbb{Z} \oplus \mathbb{Z} \oplus \ldots \oplus \mathbb{Z}$ actions.

Let M be a C^∞ manifold and $\mathrm{Diff}^r(M)$ be the set of C^r diffeomorphisms of M with the C^r Whitney topology for $r \geq 1$. A generic (second category) subset of $\mathrm{Diff}^r(M)$ is one that contains the intersection of countably many open and dense subsets of $\mathrm{Diff}^r(M)$. A $C^{s,r}$ flow on M is a C^s homomorphism of groups $\varphi\colon \mathbb{R} \to \mathrm{Diff}^r(M)$, where the group structure in $\mathbb{R}$ is given by addition and in $\mathrm{Diff}^r(M)$ by composition. When $s = 0$ we call φ a continuous flow of C^r diffeomorphisms. Except for a brief mention, we will not consider here the case $s = r = 0$ which corresponds to topological flows.

An old question is the following: when does a given $f \in \mathrm{Diff}^r(M)$ embed in a $C^{s,r}$ flow φ (i.e. $\varphi(1) = f$)? Concerning the embedding in topological flows we only mention that one can give a precise answer for generic C^1 diffeomorphisms of a compact two-dimensional manifold [8]. For the real line, Bödewadt [2] showed that C^r diffeomorphisms without fixed points embed in many C^r flows. For the circle S^1 there are two kinds of "opposite" results. On one hand, it follows from a remarkable result of Herman [4,16] that among the C^∞ diffeomorphisms of S^1 with irrational rotation number there is a dense subset whose elements embed in C^∞ flows. However, for an open and dense subset of $\mathrm{Diff}^r(S^1)$, $r \geq 2$, Kopell [5] and Lam [6] showed that no element embeds in a continuous flow of C^r diffeomorphisms. Recently, Sergeraert [13] proved that a local C^∞ flat contraction at $0 \in \mathbb{R}$ embeds in a C^1 flow but not always in a C^r flow for $r \geq 2$. For a general compact m-dimensional manifold M without boundary, we have [9]

<u>Theorem</u> - There is a generic subset G of $\mathrm{Diff}^1(M)$ such that if $f \in G$ then f does not embed in a continuous flow of diffeomorphisms.

The obstruction to prove the same theorem for $\mathrm{Diff}^r(M)$ with $r \geq 2$ [3], [9] is Pugh's Closing Lemma [10] which is known to be true only for the C^1 topology. So it is in a sense natural to look for similar and even stronger results restricted to special but important subsets of $\mathrm{Diff}^r(M)$. To do this we first set some definitions. We will consider M to be compact without boundary.

Let $f \in \mathrm{Diff}^r(M)$. A point $x \in M$ is nonwandering if for any neighborhood U of x and any integer $n_o > 0$ there is an integer $n > n_o$ such that $f^n U \cap U \neq \phi$. The set of nonwandering points is denoted by $\Omega(f)$; $\Omega(f)$ is closed and invariant, i.e. if $x \in \Omega(f)$ then the orbit $\{f^n(x); n \in \mathbb{Z}\} \subset \Omega(f)$. We say that $\Omega(f)$ is hyperbolic if

(i) the tangent bundle of M restricted to $\Omega = \Omega(f)$ can be written as a continuous direct sum of two subbundles $T_\Omega M = E^s \oplus E^u$ which are invariant by the derivative df of f,

(ii) there is a riemannian metric on M and a constant $0 < \lambda < 1$ such that for $x \in \Omega$, $v \in E^s_x$, $u \in E^u_x$

$$\|df_x\, v\| \leq \lambda \|v\|, \quad \|df_x^{-1}\, u\| \leq \lambda \|u\|.$$

In this case, the sets $W_x^s = \{z \in M; \ d(f^n x, f^n z) \to 0 \ \text{as} \ n \to \infty\}$ and $W_x^u = \{z \in M; \ d(f^{-n}x, f^{-n}z) \to 0 \ \text{as} \ n \to \infty\}$ are immersed submanifolds of M called the stable and unstable manifolds of $x \in \Omega$. Here d denotes the distance induced by the riemannian metric. We say that f satisfies Axiom A [14] if $\Omega(f)$ is hyperbolic and $\Omega(f) =$ Closure $\mathrm{Per}(f)$, where $\mathrm{Per}(f)$ is the set of periodic orbits of f. Finally we impose a "strong" transversality condition: for any pair $x,y \in \Omega(f)$, W_x^s and W_y^u are transverse. Let $\mathcal{G} \subset \mathrm{Diff}^r(M)$ be the set of diffeomorphisms satisfying Axiom A and the strong transversality condition. When $f \in \mathcal{G}$ and $\Omega(f)$ is finite, we call f a Morse-Smale diffeomorphism. Another important case is when we can extend the hyperbolic structure defined above for $\Omega(f)$ to all of M; such an f is called Anosov. Thus $\mathcal{G}$ contains many interesting examples of diffeomorphisms and in general $\mathcal{G}$ is a nonempty open subset of $\mathrm{Diff}^r(M)$.

Let us now return to the embedding problem for elements of $\mathcal{G}$. Notice that if $f \in \mathrm{Diff}^r(M)$ embeds in a continuous flow of C^r diffeomorphisms then we have a one parameter family of diffeomorphisms commuting with f. So the following concept is relevant in this context. The centralizer of $f \in \mathrm{Diff}^r(M)$ is defined as the set $C(f) = \{g \in \mathrm{Diff}^r(M); \ gf = fg\}$. Of course $C(f)$ contains the identity and it is a group under composition. When $C(f)$ is discrete then in particular f does not embed in a flow of C^r diffeomorphisms. For the circle S^1, Kopell [5] showed that there is an open and dense subset of $\mathrm{Diff}^r(S^1)$, $r \geq 2$, whose elements have trivial centralizers. We say that f has a trivial centralizer if $C(f) = \{f^n; \ n \in Z\}$. For C^∞ diffeomorphisms of a higher dimensional manifold M, we have the following

<u>Theorem</u> - There is an open and dense subset of $\mathcal{G} \subset \mathrm{Diff}^\infty(M)$ whose

elements have discrete centralizers. That is, among the C^∞ diffeomorphisms of M satisfying Axiom A and the strong transversality condition the ones with discrete centralizers contain an open and dense subset.

Sketch of the Proof: Given any $f \in \mathcal{C}$, we show that there exist $f_1 \in \mathcal{C}$ near f and a neighborhood U of f_1 in $\mathcal{C}$ such that every $f_2 \in U$ has a discrete centralizer. To do this we first consider the spectral decomposition [14] of $\Omega(f) = \Omega_1 \cup \Omega_2 \cup \ldots \cup \Omega_k$. Each Ω_i is called a basic set and it is closed, invariant and transitive for f and $\mathrm{Per}(f) \cap \Omega_i$ is dense in Ω_i. Some of them are attractors, some repellors and the remaining are of saddle type. An important fact is that the stable and unstable manifolds of any periodic orbit in Ω_i are dense in Ω_i. From this it follows that if h is a homeomorphism near the identity 1 on M and $hf = fh$ then $h = 1$ on each Ω_i. Now we observe that the union of the stable manifolds of the attractors is dense in M and the same is true for the union of the unstable manifolds of the repellors. So we can restrict our attention to the stable manifolds of the attractors and the unstable manifolds of the repellors. Moreover, since the stable manifold of any periodic orbit in Ω_i is dense in the stable manifold of Ω_i and similarly for the unstable manifolds, we can further restrict the argument to a finite number of periodic orbits, one in each attractor and each repellor. Let Ω_1 be an attractor, Ω_k a repellor such that $W^s(\Omega_1) \cap W^u(\Omega_k) \neq \emptyset$ and let $p_1 \in \Omega_1$, $p_k \in \Omega_k$ be periodic orbits. We can perturb f if necessary so that $f/W^s(p_1)$ and $f/W^u(p_k)$ are locally linearizable. Thus any $h \in \mathrm{Diff}^\infty(M)$ near 1 and $hf = fh$ must leave $W^s(p_1)$ and $W^u(p_k)$ invariant and moreover $h/W^s(p_1)$ and $h/W^u(p_k)$ must have a very special form that depends only on the

the germs of f near p_1 and p_k [5]. On the other hand h must also leave invariant the local connected components of $W^s(p_1) \cap \cap W^u(p_k)$. Thus we can further perturb f outside Ω_1 and Ω_k to f_1 so that if $hf_1 = f_1 h$ and h is near 1 then h must be 1 on $W^s(p_1)$ and consequently on all of $W^s(\Omega_1)$. We repeat the argument a finite number of times for pairs attractor-repellor. All the steps work as well for f_2 very near f_1 and so we obtain the desired neighborhood U whose elements have discrete centralizer.

This theorem generalizes Walters [15] for Anosov diffeomorphisms and Anderson [1] for Morse-Smale diffeomorphisms.

We now put together the above result with a basic fact concerning strucutral stability of diffeomorphisms in order to exhibit many examples of structurally stable foliations. This idea was explored in [1], [7].

A diffeomorphism f is structurally stable if for any diffeomorphism f^* C^r close to f there exists a homeomorphism $h: M \to M$ such that $hf = f^*h$. In particular h sends the orbits of f onto the orbits of f^*. From [11], [12] if $f \in \mathcal{G}$ then f is structurally stable. Moreover, it is conjectured that f is structurally stable iff $f \in \mathcal{G}$.

Let us recall briefly the construction of a suspended foliation. Some of our assumptions can be removed without trouble, but we will not worry about this here. Let M, N be C^∞ compact manifolds without boundaries and let $\pi_1(N)$ the fundamental group of N, be an infinite group finitely generated. From a representation (group homomorphism) ρ of $G = \pi_1(N)$ into $\mathrm{Diff}^\infty(M)$ we can obtain a C^∞ foliation on an M-bundle over N whose leaves are transverse to the fiber M. This is done as follows. The representation ρ induces a free action γ of G on $\tilde{N} \times M$, where $\tilde{N}$ is the universal covering space of N: $\gamma(g)(n,m) = (ng, \rho(g)^{-1} m)$, the action on the first factor being by deck transformations. From this we get that $\tilde{N}\times M/G \to N$ is an M-bundle over N and since γ preserves the trivial foliation $\tilde{N}\times M \to M$ it induces a foliation on $\tilde{N}\times M/G$

whose leaves are transverse to the fiber M.

The set of representation of G into $\mathrm{Diff}^\infty(M)$ inherits naturally a topology from that of $\mathrm{Diff}^\infty(M)$. So two representations $\rho, \rho^*: G \to \mathrm{Diff}^\infty(M)$ are close if $\rho(g_i)$ is close to $\rho^*(g_i)$ for $1 \le i \le k$ where $g_1, g_2, \ldots, g_k$ is a basis for G. A representation ρ is structurally stable if for any ρ^* near ρ there is a homeomorphism $h: M \to M$ such that $h\rho(g_i) = \rho^*(g_i)h$ for all $1 \le i \le k$. On the other hand the C^∞ foliations of $N_1 = \tilde{N} \times M/G$ can be topologized using local trivializations. A foliation $\mathfrak{F}$ of N_1 is structurally stable if for any nearby foliation $\mathfrak{F}^*$ there is a homeomorphism $h: N_1 \to N_1$ sending leaves of $\mathfrak{F}$ onto leaves of $\mathfrak{F}^*$. It is not hard to show the following

Proposition - The representation $\rho: G \to \mathrm{Diff}^\infty(M)$ is structurally stable iff its induced foliation $\mathfrak{F}_\rho$ is structurally stable.

Let now $G = \pi_1(N)$ be abelian and consider the special representations $\rho(g_1) = f$ and $\rho(g_i) = 1$ for $2 \le i \le k$ where 1 means the identity on M.

Proposition - The representation $\rho: G \to \mathrm{Diff}^\infty(M)$ given by $\rho(g_1) = f$ and $\rho(g_i) = 1$ for $2 \le i \le k$ is structurally stable iff

(1) f is structurally stable,

(2) there is a neighborhood $V(f) \subset \mathrm{Diff}^\infty(M)$ such that if $f^* \in V$ then $C(f^*)$ is discrete.

Thus our main theorem and these two propositions provide many examples of structurally stable foliations. Moreover, modulo the conjecture "if the diffeomorphism f is structurally stable then $f \in \mathcal{G}$", we provided a characterization of the structurally stable foliations induced by representations of the form $\rho(g_1) = f$, $\rho(g_2) = \ldots = \rho(g_k) = 1$.

A very beautiful question is to look for a similar characterization for general representations $\rho(g_1) = f_1$, $\rho(g_2) = f_2, \ldots, \rho(g_k) = f_k$. The following question is interesting in itself and it may be relevant to the previous one: is it true for an open and dense subset of G that its elements have trivial centralizers?

References

[1] R.B. Anderson, The centralizer of a Morse-Smale diffeomorphism, Berkeley thesis (1973).

[2] U.T. Bödewadt, Zur Iteration reeler Funktionen, Math. Z. 49 (1944), 497-516.

[3] M.I. Brin, On embedding a diffeomorphism in a flow, Izvestia Mat. 123 (1972), 19-25.

[4] M. Herman, Les difféomorphismes du cercle, to appear.

[5] N. Kopell, Commuting diffeomorphisms, Proc. Symp. Pure Math., Vol. 14, Amer. Math. Soc. (1970), 165-184.

[6] P.F. Lam, Embedding a homeomorphism in a flow subject to differentiability conditions, Topological Dynamics, Benjamin - New York (1968), 319-333.

[7] H. Levine and M. Shub, Stability of foliations, Trans. Amer. Math. Soc. 184 (1973), 419-437.

[8] J. Palis, On Morse-Smale dynamical systems, Topology 8 (1969), 385-404.

[9] J. Palis, Vector fields generate few diffeomorphisms, Bull. Amer. Math. Soc. 80 (1973), 503-505.

[10] C. Pugh, An improved closing lemma and a general density theorem, Amer. J. Math. 89 (1967), 1010-1021.

[11] J. Robbin, A structural stability theorem, Annals of Math. 94 (1971), 447-493.

[12] C. Robinson, Structural stability for C^1 diffeomorphisms, Dynamical Systems-Warwick, Springer-Verlag (1975), 21-23.

[13] F. Sergeraert, Plonger les difféomorphismes dans les flots, to appear.

[14] S. Smale, Differentiable dynamical systems, Bull. Amer. Math. Soc. 73 (1967), 747-817.

[15] P. Walters, Anosov diffeomorphisms are topologically stable, Topology 9 (1970), 71-78.

[16] H. Rosenberg, Some remarks on the Arnold conjecture and the theorem of Michel Herman , these Proceedings.

Integrable Perturbations of Fibrations and a theorem of Seifert

Remi Langevin and Harold Rosenberg

Let $p: E \to B$ be a smooth fibration with fibre F. The problem we wish to consider concerns foliations F whose plane fields are close, in some C^r-topology, to the plane field F_o tangent to the fibres. We shall always assume E,F,B are closed manifolds. A natural and provoking question is when does F have a compact leaf? The first result of this nature we know of is due to H. Seifert [5]. He proved that any C^o perturbation of the Hopf fibration $S^3 \to S^2$ has a compact leaf. In the same paper he announced that the theorem is also true for orientable S^1 bundles over surfaces B of $x(B) \neq 0$; $x(B)$ is the euler characteristic of B. This result was generalised by F. Fuller to orientable circle bundles over arbitrary closed manifolds B with $x(B) \neq 0$, [1].

We have proved a fibration F_o is C^1 structurally stable if and only if $H^1(F,R) = 0$, [2]. In particular, if $\Pi_1(F)$ is finite, then any perturbation of F_o has all compact leaves and is a fibration equivalent to F_o. Now we shall consider fibrations with $\Pi_1(F) \approx Z$. We consider the cases dimension B equals one and two. When B is a surface with $x(B) \neq 0$, and $\Pi_1(B)$ operates trivially on $\Pi_1(F)$, we shall prove any C^o-perturbation of F_o has a compact leaf. When $B = S^1$ and the monodromy is multiplication by -1 on $\Pi_1(F)$ we also prove any perturbation of F_o has a compact leaf. The reader should first convince himself this is true for the Klein bottle (in fact, any foliation of the Klein bottle has a compact leaf, but that is not the proper point of view here). We would like to thank David Epstein for stimulating conversations.

§1. <u>The First Return Map</u>. Henceforth, we assume $\Pi_1(F)$ is isomorphic to Z and $\Pi_1(B)$ operates trivially on $H_1(F)$, hence on $\Pi_1(F)$. We let $\mathcal{F}$ be a sufficiently small perturbation of the fibration $\mathcal{F}_0$. We shall associate to $\mathcal{F}$ a diffeomorphism $f:E \to E$ (a generalised first return map for vector field perturbations of orientable circle bundles) and if x is a fixed point of f, the leaf of $\mathcal{F}$ through x is compact and isotopic to the fibre F(x) of $\mathcal{F}_0$ through x by an isotopy contained in some tubular neighbourhood of F(x). In particular, if one knows that f has a fixed point then $\mathcal{F}$ has a compact leaf.

Fix a Riemannian metric on E and choose $\varepsilon > 0$ so that for each $x \in E$, the geodesics through x, of length ε, and orthogonal to F(x), form a smoothly embedded disc D(x). We can suppose that for each fibre F, the discs D(x), $x \in F$, form a tubular neighbourhood of F which we denote by T(F).

Fix a fibre F and $x \in F$. Let α be a loop in F at x representing a generator of $\Pi_1(F)$. Then for $\mathcal{F}$ close to $\mathcal{F}_0$, α can be lifted to a path on the leaf of $\mathcal{F}$ through x, to a path starting at x and ending at a point of D(x). This endpoint is denoted by $H(\mathcal{F},\alpha)(x)$: H is the perturbed holomony map (cf. [2] and [3] for details). We define $f(x) = H(\mathcal{F},\alpha)(x)$. Now if y is another point of F(x), let β be any path in F from x to y (the length of β less than the diameter of F) and define $f(y) = H(\mathcal{F},\beta\alpha\beta^{-1})(y)$. This does not depend on β and defines a smooth map $f:F(x) \to E$. For each $y \in F(x)$, we have $f(y) \in D(y)$ and if $f(y) = y$ then the leaf of $\mathcal{F}$ through y is compact ([2]). Next, one extends f to a map f: $T(F(x)) \to E$ by using the product structure in $T(F(x))$ and transporting α to each fibre in $T(F(x))$.

Now extend f to E by using chains of trivialised product neighbourhoods $T_1, T_2, \ldots, T_k$ with $T_i \cap T_{i+1} \neq \emptyset$ for $i = 1, \ldots, k-1$. Since $\Pi_1(B)$ acts trivially on $\Pi_1(F)$ this does not depend on the chain and gives a well defined map $f: E \to E$.

We associate to $\mathcal{F}$ a vector field X whose zero's give compact leaves. We have x and f(x) in the geodesic disc D(x) for each $x \in E$. Let X(x) be the vector tangent to the geodesic in D(x) from x to f(x). If f has no fixed points then X(x) is never zero and is orthogonal to $\mathcal{F}_0$.

§2. Perturbations in codimension two. Let B be a closed two manifold and $p: E \to B$ a fibration with $\Pi_1(F) = Z$ and $\Pi_1(B)$ acting trivially on $\Pi_1(F)$. Let D be a disc in B over which E is trivial. We identify D with the unit disc $D^2 \subset R^2$ and $p^{-1}(D) = T$ with $D \times F$. Let $\mathcal{F}$ be a small perturbation of $\mathcal{F}_0$ so that the first return map f and vector field X are defined as in §1.

Let $h: S^1 \to T$ be a continuous map, $h(S^1) = C$. We define $I(\mathcal{F}, C)$ to be the total number of times the vector $p_*(X(h(\theta)))$ rotates about the origin of D as θ goes once around S^1; it's the degree of the map $S^1 \to R^2 - \{0\}$, $\theta \to p_*(X(h(\theta)))$.
This makes sense provided X is not zero along C and depends only on the homotopy class of C in T.

Let $S_1 \subset E$ be a smooth section over B-(int D), meeting ∂T in a simple closed curve C_1. The vector field X projects naturally to a vector field $\tilde{X}$ tangent to S_1 since X and S_1 are transverse to $\mathcal{F}_0$. Then $Y = p_*(\tilde{X})$ is a vector field on Σ^2 = B-int D. If $\mathcal{F}$

has no compact leaves then Y is non zero, and the obstruction to extending Y to a non zero field on D is $\chi(B)$ and this is the total number of times Y turns as we traverse ∂D once.

Let $F_0 = p^{-1}(0)$ where $0 \in D$ is the origin. Let α be a loop in F_0 representing a generator of $\Pi_1(F_0)$. Let i be the integer defined by $[C_1] = i[\alpha]$, where $[\ ]$ denotes homotopy class in $\Pi_1(T)$. If $i = 0$ then E admits a section S and $\mathfrak{X}$ is a non zero vector field on S so $\chi(B) = 0$. In any case, we always have:

$$\chi(B) = i.I(\mathcal{F},\alpha).$$

Now Seifert has proved $I(\mathcal{F},\alpha) = 0$ for orientable circle bundles and his proof is local in the following sense.

Lemma 2.1 (Seifert [5]). Let n be a vector field on $S^1 \times D^2$ which is a C^0-perturbation of the vector field $n_0 = (1,0)$. Suppose that n has no compact orbits on $S^1 \times (\frac{1}{2}D)$, where $\frac{1}{2}D = \{x \in R^2/\|x\| < \frac{1}{2}\}$. Let $\alpha = S^1 \times \{0\}$. Then $I(n,\alpha) = 0$.

Now we can prove:

Theorem 2.2 Let $p: E \to B$ be a smooth fibration with fibre F, B a closed two manifold, E closed. Suppose:

1) $\Pi_1(F) = Z$
2) $\Pi_1(B)$ acts trivially on $\Pi_1(F)$
3) $\chi(B) \neq 0$.

Then any foliation $\mathcal{F}$ of E C^0-close to the fibration has a compact leaf.

Proof. It suffices to prove $I(\mathcal{F},\alpha) = 0$. Consider the solid torus $L = \alpha \times D$. The fibration is transverse to L and induces the one dimensional foliation by circles $\alpha \times \{pt\}$. Thus $\mathcal{F}$ is transverse to L and induces a one dimensional foliation G. If we assume $\mathcal{F}$ has no compact leaves then G has no compact leaves since

$$H(G)(\alpha)(x) = H(\mathcal{F})(\alpha)(x),$$

for $x \in L$. Clearly $I(G,\alpha) = I(\mathcal{F},\alpha)$, so by Seifert's Lemma 2.1, $I(\mathcal{F},\alpha) = 0$ hence $x(B) = 0$ which contradicts the hypothesis.

§3. Bundles over S^1. Let $p:E \to S^1$ be a fibration with fibre F. Let $h:F \to F$ be the monodromy diffeomorphism; E is obtained from $F \times [0,1]$ by identifying $(x,1)$ with $(h(x),0), x \in F$.

Theorem 3.1. If $\Pi_1(F) = Z$ and h is multiplication by -1 on $\Pi_1(F)$ then every foliation C^0-close to the fibration has a compact leaf.

Proof. We can suppose h has a fixed point z. Let Z be a vector field transverse to the fibration, with integral curves ϕ_t leaving $\mathcal{F}_o$ invariant such that $\phi_1 = h$. Let $C = \{\phi_t(z)/0 \leq t \leq 1\}$. C is a simple closed curve, transverse to $\mathcal{F}_o$, meeting each fibre once. Let $\alpha(0)$ be a loop at z in $F(z)$ representing a generator of $\Pi_1(F(z))$. Let $\alpha(t)$ be the loop in $F(\phi_t(z))$ defined by $\phi_t(\alpha(0))$.

Since $\phi_1 = h$ we have $[\alpha(1)] = -[\alpha(0)]$ in $\Pi_1(F(0))$. Suppose there exist perturbations $\mathcal{F}$ of $\mathcal{F}_o$ having no compact leaves. Then for $0 \leq t \leq 1$, we have

$$H(\mathcal{F},\alpha(t))(\phi_t(z)) \neq \phi_t(z).$$

Let $Y(t)$ be the non zero vector joining $\phi_t(z)$ to $H(\mathcal{F},\alpha(t))(\phi_t(z))$. Clearly $Y(t)$ is transverse to $\mathcal{F}_o$, and $Y(1) = -Y(0)$. But this is impossible; e.g., the scalar product of $Y(t)$ and $Z(\phi_t(z))$ would have to change sign. Hence $\mathcal{F}$ has a compact leaf.

University of Paris - Orsay
University of Warwick, Coventry

April, 1976.

BIBLIOGRAPHY

1. F. Fuller. An index of fixed point type for periodic orbits. American Journal of Maths. 1967 (89) pp 133-148.
2. R. Langevin and H. Rosenberg. On stability of compact leaves and fibrations, Topology 16 (1977), 107-112".
3. M. Hirsch. Stability of compact leaves of foliations. Dynamical Systems, Academic press, pp 135-155, 1971.
4. G. Reeb. Sur un théorème de Seifert sur les trajectoires fermées de certains champs de vecteurs. International symposium on non linear differential equations and non linear mechanics, 1963.
5. H. Seifert. Closed integral curves in 3-space and isotopic two dimensional deformations. Proc. A.M.S., pp 287-302, 1950.

Structural Stability of Foliations with Singularities

by

César Camacho

Consider a differential 1-form ω satisfying the integrability condition $\omega \wedge d\omega = 0$ on a manifold M. A singularity of ω is a point $x \in M$ where $\omega_x = 0$. Let $\mathrm{Sing}(\omega)$ be the set of singular points of ω.

Let ω, η be integrable forms on M. We say that ω and η are topologically equivalent if there is a homeomorphism $h: M \to M$ such that (1) $h(\mathrm{Sing}(\omega)) = \mathrm{Sing}(\eta)$, (2) h sends leaves of ω onto leaves of η.

The form ω is called C^r-structurally stable if there is a neighborhood $N(\omega)$ in the uniform C^r-topology such that any $\eta \in N(\omega)$ is topologically equivalent to ω.

The central problem concerning the stability of forms is to characterize the integrable forms which are structurally stable. Regular foliations of dimension greater than one are rarely structurally stable on some manifolds. For instance no foliation of codimension one of S^3 is C^1-structurally stable ([10]). We consider here only foliations with singularities i.e. those defined by forms ω such that $\mathrm{Sing}(\omega) \neq \emptyset$.

§1. Local Theorems

Historically, the first to investigate the local structure of integrable forms near a singularity was G. Reeb - and from the point of view of structural stability, I. Kupka. -

THEOREM (Reeb [8]) - Let ω be a C^r integrable form, $r \geq 2$, in R^n, $\omega_0 = 0$, and $\omega^1 = J^1[\omega]_0$ the 1-jet of ω at $0 \in \mathbb{R}^n$. Then ω^1 is integrable and

(i) If $d\omega^1 = 0$ there is a linear change of coordinates A such that $A^*\omega^1 = \sum^n \epsilon_i x_i dx_i = df$ where $\epsilon_i = \pm 1$ or 0. If f has index 0 or n then ω admits a C^{r-1} first integral. Moreover ω is locally equivalent to ω^1.

(ii) If $d\omega^1 = 0$ and ω is analytic with ω^1 non degenerate, Then ω admits an analytic first integral and it is locally equivalent to ω^1.

(iii) If $d\omega_1 \neq 0$ then there is a linear change of coordinates A such that $A^*\omega^1 = L_1(x_1,x_2)\, dx_1 + L_2(x_1,x_2)\, dx_2$ where the L_i are linear functions.

In [3] Kupka announced the following theorem.

THEOREM - Let ω be an integrable form in a compact manifold M. Suppose that ω is C^1-structurally stable and let $x^o \in \mathrm{Sing}(\omega)$. If $d\omega_{x^o} = 0$ there is a neighborhood $V \ni x^o$ and maps $f,g: V \to \mathbb{R}$, f with a unique critical point x^o which is generic and $g(x) \neq 0$ for any $x \in V$ such that $\omega = gdf$. If $d\omega_{x^o} \neq 0$ there exist a neighborhood $N_o \ni x^o$, a C^2 map $\varphi_o: N_o \to \mathbb{R}^2$ of rank 2 and a unit $g_o: N_o \to \mathbb{R}$ such that if $\alpha_o = \eta d\xi - \lambda\xi d\eta$, $\lambda \neq 0$, 1 scalar, and $\alpha_1 = \xi d\xi + \eta\, d\eta + \mu(\eta d\xi - \xi d\eta)$, $\mu \neq 0$ scalar, denote forms on $\mathbb{R}^2$ then either $\omega/N_o = g_o\, \varphi_o^*\, \alpha_o$ or $\omega/N_o = g_o\, \varphi_o^*\, \alpha_1$.

In [9] (page 27) Reeb asked for a better understanding of the case $d\omega^1 \neq 0$. An answer to this is given in the next theorem in a more general context for p-forms. Sufficient conditions for structural stability are also given there.

THEOREM (A. Medeiros [5])

(1) Let ω be in $\mathbb{R}^n$, $\omega_o = 0$, $\omega \wedge d\omega \equiv 0$ and $d\omega_o = 0$. Assume that the linear part of ω at $0 \in \mathbb{R}^n$, $J^1[\omega]_o$ is non degenerate

with index $\neq 2$, $n-2$. Then ω is locally C^1-structurally stable at $0 \in \mathbb{R}^n$ and ω is topologically equivalent to $J^1[\omega]_o$ in a neighborhood of $0 \in \mathbb{R}^n$.

(2) Let Ω be a p-form in $\mathbb{R}^n$, $p \geq 1$, $\Omega_o = 0$, $d\Omega_o \neq 0$. Assume Ω is of class C^r, $r \geq 2$, and for any $x \notin \text{Sing}(\Omega)$ near $0 \in \mathbb{R}^n$ there is a neighborhood $V_x \ni x$ and 1-forms $\eta_1,\dots,\eta_p$ in V_x such that $\Omega = \eta_1 \wedge \dots \wedge \eta_p$ and the system $\{\eta_1,\dots,\eta_p\}$ is integrable. Then there is a neighborhood $U \ni 0$ and a C^{r-1} change of coordinates $f\colon U \to \mathbb{R}^n$, $f(0) = 0$, such that $f^*(\Omega)$ depends only on $p+1$ variables.

$$f^*(\Omega) = \sum_{1\leq i_1<\dots<i_p\leq p+1} a_{i_1\dots i_p}(x_1,\dots,x_{p+1})\, dx_{i_1} \wedge \dots \wedge dx_{i_p}.$$

Moreover, Ω is C^{r-2} structurally stable in a neighborhood of $0 \in \mathbb{R}^n$ if and only if the vector field Y defined by $\Omega = i_Y(dx^1 \wedge \dots \wedge dx^{p+1})$ has a hyperbolic singularity at $0 \in \mathbb{R}^{p+1}$. In this case Ω is topologically equivalent to $J^1[\Omega]_o$.

REMARK - If ω is an integrable form in $\mathbb{R}^n$ with linear coefficients $\omega_o = 0$, $d\omega_o = 0$, then it is easy to see after using (1), that a necessary and sufficient condition of C^r-structural stability of ω is that ω is nondegenerate of index $\neq 2$, $n-2$. However when ω is non linear this question is still open. Presumably the work of R. Moussu [6] will lead to an answer for this.

A natural question arises here and this is:
Is it possible to approximate in the C^r-topology any integrable form ω, $\omega_o = 0$ by another η, integrable, $\eta_o = 0$ such that $J^1[\eta]$ is structurally stable?
If $d\omega_o \neq 0$ the answer is yes and it follows from the theorem above.
If $d\omega_o = 0$ the answer is not always if $r \geq 2$ as we can see from the following example in R^3:

$$q = \lambda_1 \, x_2 \, x_3 \, dx_1 + \lambda_2 \, x_1 \, x_2 \, dx_2 + \lambda_3 \, x_1 \, x_2 \, dx_3$$

$$\lambda_i \neq \lambda_j \quad \text{for} \quad i \neq j.$$

Then $q \wedge dq = 0$ and $X = \mathrm{rot}(q)$ defined as $dq = i_X(dx^1 \wedge dx^2 \wedge dx^3)$ is a linear hyperbolic vector field in $\mathbb{R}^3$. We claim that for any η with $\eta \wedge d\eta \equiv 0$, $\eta_o = 0$, C^2-close to q one must have $J^1[\eta]_o \equiv 0$. This is because $\mathrm{rot}(\eta)$, being C^1 close to $\mathrm{rot}(q)$, has a hyperbolic singularity at $0 \in R^n$. As $d\eta_o = 0$, $J^1[\eta] = df$. The condition $\eta \wedge d\eta \equiv 0$ implies that f is constant along the orbits of the linear part of $\mathrm{rot}(\eta)$. Thus $df \equiv 0$. This motivates the study of forms q with quadratic coefficients such that $\mathrm{rot}(q)$ is hyperbolic.

<u>Definition</u> - Let q be a 1-form in $\mathbb{R}^3$ with quadratic coefficients and $q \wedge dq = 0$. We say that q is hyperbolic if it can be written in some system of coordinates (x_1, x_2, x_3) in one of the forms:

(a) $q = \lambda_1 \, x_2 \, x_3 \, dx_1 + \lambda_2 \, x_1 \, x_3 \, dx_2 + \lambda_3 \, x_1 \, x_2 \, dx_3$

$$\lambda_i \neq \lambda_j \quad \text{for} \quad i \neq j.$$

(b) $q = (\alpha x_1 + \beta x_2) \, x_3 \, dx_1 + (-\beta x_1 + \alpha x_2) \, x_3 \, dx_2 + \gamma(x_1^2 + x_2^2) \, dx_3$

$$\alpha, \beta, \gamma \neq 0.$$

THEOREM (A. Lins Neto [4]) - Let ω be a 1-form in R^3, $\omega \wedge d\omega = 0$, $\omega_o = 0$, $d\omega_o = 0$ and such that $J^1[\omega]_o = 0$ and $J^2[\omega]_o$ is hyperbolic. Then ω is C^2-structurally stable and it is topologically equivalent to $J^2[\omega]_o$.

There are three topological types for ω as above:

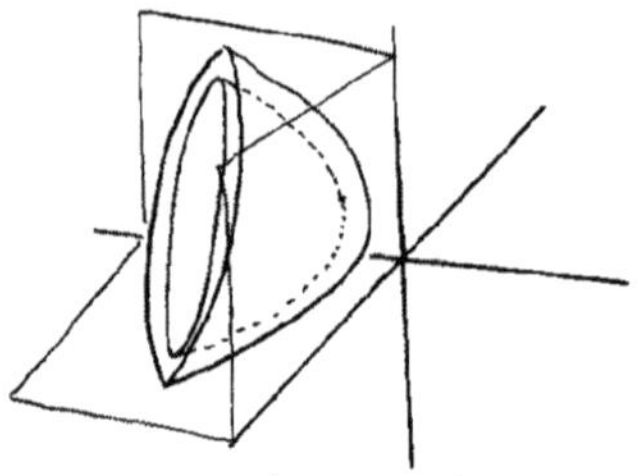

$\mathrm{Sing}(\omega) = \{x = (x_1, x_2, x_3) \mid x_i x_j = 0 \ i \neq j\}$

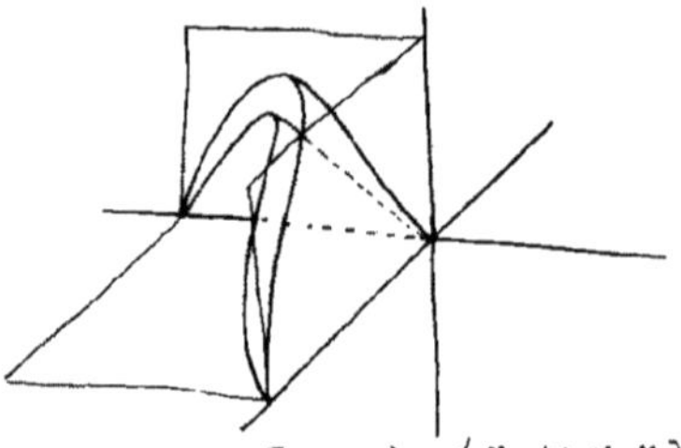

$\mathrm{Sing}(\omega) = \{x = (x_1, x_2, x_3) \mid x_i x_j = 0 \ i \neq j\}$

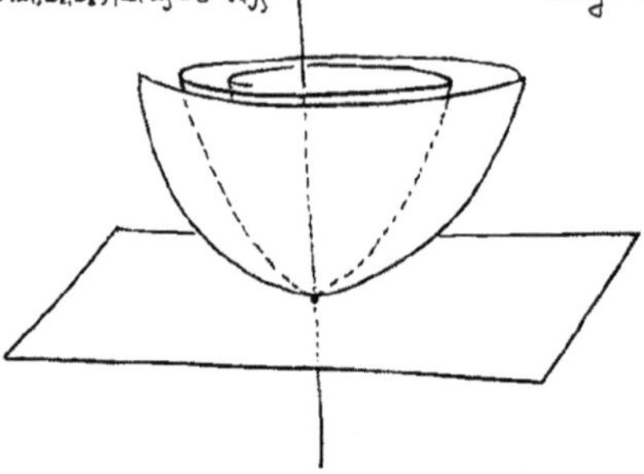

$\mathrm{Sing}(\omega) = \{x = (x_1, x_2, x_3) \mid x_1 = x_2 = 0\}$.

REMARK - The leaves of $q = J^2[\omega]$ above, are the orbits of a hyperbolic linear action of the group $\mathbb{R}^2$ on $\mathbb{R}^3$ (see [1]).

There is a generalization of the above theorem to $\mathbb{R}^n$ and it goes as follows ([4]): If ω is integrable in $\mathbb{R}^n$, $\omega_0 = 0$, $J^1[\omega]_0 \equiv 0$ and $J^2[\omega]_0$ is quadratic hyperbolic in a subspace $\pi \subset \mathbb{R}^n$ of dimension three then there is a change of variables φ of class C^{r-3} in the neighborhood of zero such that $\varphi^*(\omega)$ depends only on three variables. These forms are also equivalent to actions of $\mathbb{R}^{n-1}$ on $\mathbb{R}^n$ in the neighborhood of a hyperbolic orbit of dimension $n-3$.

Consider a submanifold $S \subset M$ such that $\omega_x = 0$ and $d\omega_x \neq 0$ for any $x \in S$.

Definition - S is called normally hyperbolic if for some $x^0 \in S$ there is a system of coordinates $(x_1, \ldots, x_n)$ in a neighborhood U of x^0 such that $\omega = a_1(x_1, x_2)dx_1 + a_2(x_1, x_2)dx_2$ and the vector field $Y^{\omega} = (-a_2, a_1)$ has a hyperbolic singularity at x^o.

The form $d\omega$ defines a foliation $\mathfrak{F}$ by $\{X;\ i_X(d\omega) = 0\}$. This foliation has codimension two is tangent to the leaves of ω, and has S as one of its leaves. Therefore the definition above is independent of the point $x^o \in S$.

Call $\Sigma_o = \{(x_1, x_2, 0, \ldots, 0) \in U\}$. Then the orbits of Y^{ω} on Σ_o are the intersections of the leaves of ω with Σ_o. Assume $\pi_1(S) = Z$ and let $f\colon V \to \Sigma_o$, $x_o \in V \subset \Sigma_o$, be the holonomy map of $\mathfrak{F}$. Then f is a diffeomorphism sending orbits of Y^{ω} to orbits of Y^{ω} and $f(x^o) = x^o$.

Definition - Assume S is compact. Then S is called hyperbolic if there is a splitting $T_{x^o}\Sigma_o = E_1 \oplus E_2$ invariant by $Df(x^o)$ and $DY^{\omega}(x^o)$ and if $(\lambda_i)_{i=1}^2$ $(\mu_i)_{i=1}^2$ denote the eigenvalues of $DY^{\omega}(x^o)$ and $Df(x^o)$ then $\lambda_1 \log|\mu_2| \neq \lambda_2 \log|\mu_1|$.

The following theorem yields the local structure of ω in the neighborhood of S.

THEOREM ([5]) - Let ω be an integrable 1-form on a manifold M and $S \subset M$, $\pi_1(S) = \mathbb{Z}$, a compact submanifold of singularities of ω such that $d\omega_x \neq 0$ for $x \in S$. Assume S is hyperbolic for some $g\omega$, $g \neq 0$ everywhere. Then ω is C^r-structurally stable in a neighborhood of S.

§2. Global Theorems

Let ω be an integrable 1-form on a compact 3-manifold M.

A limit point of a non compact leaf F of ω is an accumulation point in M of a divergent sequence in F. The union of all these points will be denoted $\lim(F)$. Write also $\lim(\omega) = \doteq \bigcup_F \lim(F)$.

Let x be a singular point of ω such that $d\omega_x = 0$. Then x is called <u>hyperbolic</u> if (i) $J^1[\omega]_x = df$ and f is a nondegenerate quadratic function, or (ii) $J^1[\omega]_x \equiv 0$ and $J^2[\omega]_x$ is a quadratic hyperbolic form. In (i) we say that x is a <u>cone point</u> if the index of f is one and center point otherwise.

An <u>alternate connection</u> is a leaf F such that (i) $\lim(F)$ is a closed simple curve (ii) $\lim(F) = \bigcup_{i=1}^{n} \bar{s}_i$ where each s_i is a normally hyperbolic curve, $\bar{s}_i \cap \bar{s}_{i+1} \neq \emptyset$ and $s_n = s_1$ (iii) There are at least two s_i; s_{i_1}, s_{i_3} of saddle type and two attractors s_{i_2}, s_{i_4} in alternate order i.e. $i_1 < i_2 < i_3 < i_4$.

The following Theorems are in [2].

THEOREM - Let ω be an integrable form on a compact 3-manifold M such that,

(i) All leaves of ω have abelian fundamental group and $\mathrm{Sing}(\omega)$ is a p-dimensional, $p \leq 1$, compact polyhedron whose vertices are hyperbolic singularities and 1-sides normally hyperbolic curves closed or not.

(ii) The union of $\mathrm{Sing}(\omega)$ with all nonsimply connected leaves of ω is a compact polyhedron $C(\omega)$ of dimension ≤ 2. Any leaf F in $C(\omega)$ has nontrivial linear holonomy. Moreover $\lim(F) \subset \mathrm{Sing}(\omega)$ and $\lim(\omega) \subset C(\omega)$.

(iii) There are no alternate connections.

Then ω is C^2-strutucturally stable.

THEOREM - Let ω be integrable such that any leaf of ω is simply connected. Suppose

(i) $\mathrm{Sing}(\omega)$ is a finite union of hyperbolic singular points and normally hyperbolic singular curves.

(ii) $\lim(\omega) \subset \mathrm{Sing}(\omega) \neq \emptyset$.

(iii) There are no alternate connections.

Then ω is C^2-structurally stable.

Many examples of these foliations can be constructed on S^3 and other 3-manifolds. There are examples showing that if one of the conditions of the theorems above is not fulfilled then the foliation is not stable.

§3. On $\mathbb{R}^2$-actions

Let $\varphi: \mathbb{R}^2 \times M \to M$ be an action on a compact 3-manifold M. The orbit of x $\mathcal{O}_x = \{\varphi(r,x) \mid r \in \mathbb{R}^2\}$ is called <u>singular</u> if the isotropy group $G_x = \{r \in \mathbb{R}^2 \mid \varphi(r,x) = x\}$ is non zero. A point $x \in M$ is <u>nonwandering</u> if given $n \in Z_+$ and a neighborhood $V \ni x$ there is $r \in \mathbb{R}^2$, $|r| > n$ such that $\varphi(r,V) \cap V \neq \emptyset$. Call $\Omega(\varphi)$ the set of nonwandering points of φ. Clearly any singular orbit is in $\Omega(\varphi)$. In particular any compact orbit is in $\Omega(\varphi)$.

THEOREM ([2]) - Let $\varphi: \mathbb{R}^2 \times M \to M$ be a C^1-action on a compact 3-manifold M such that

(i) The set of nonwandering points of φ is an embedded finite complex consisting of finitely many singular orbits.

(ii) Any compact orbit of φ is hyperbolic ([1]).

(iii) There are no alternate connections.

Then φ is C^1-structurally stable.

Two commuting linear vector fields on R^{n+1} induce, via the radial projection $\pi: R^{n+1}-0 \to S^n$ $\pi(x) = \frac{x}{|x|}$, an $\mathbb{R}^2$-action on S^n.

These so called linearly induced actions were introduced in [1] to get examples of $\mathbb{R}^2$-actions on spheres with all compact orbits hyperbolic.

In [7] G. Palis proves the following

THEOREM - A linearly induced $\mathbb{R}^2$-action on S^3 is structurally stable if and only if all compact orbits are hyperbolic.

REFERENCES

[1] C. Camacho, On $R^k \times Z^\ell$-actions, Dynamical Systems, Edited by M. Peixoto, 1971.

[2] C. Camacho, Structural stability of integrable differential forms on 3-manifolds. To appear.

[3] I. Kupka, The singularities of integrable structurally stable pfaffian forms, Proc. of the Nat. Acad. of Sc. Vol. 52, pg. 1431, 1964.

[4] A. Lins Neto, Local structural stability of C^2-integrable forms, To appear.

[5] A.S. Medeiros - Structural stability of integrable differential 1-forms. To appear.

[6] R. Moussu, Sur l'existence d'integrales premières pour un germe de forme de Pfaff. To appear.

[7] G. Palis, Linearly induced vector fields and R^2-actions on spheres, Dynamical Systems, Warwick 1974, Springer Lectures Notes in Mathematics nº 468.

[8] G. Reeb, Sur certaines propriétés topologiques des variétés feuilletées, Hermann, Paris 1952.

[9] G. Reeb, Feuilletages, résultats anciens et nouveaux (Painlevé Hector e Martinet) Séminaire de Mathématiques Supérieures - été 1972. Les Presses de l'Université de Montréal.

[10] H. Rosenberg, R. Roussarie, Some remarks on the stability of foliations. Jr. Diff. Geom. Vol. 10, nº 2, 1975.

UN THEOREME DE THURSTON ETABLI AU MOYEN DE L'ANALYSE NON STANDARD

par

G. REEB (Strasbourg) et P. SCHWEITZER, s. j. (Strasbourg et Rio)

Le lemme clef (voir ci-dessous) de Thurston, A Generalization of the Reeb Stability Theorem, Topology 13 (1974), admet une démonstration brève si on utilise l'analyse non standard inventée par A. Robinson. Son livre Non-Standard Analysis (la référence standard !) indique, en particulier dans la section 3.6 sur les différentielles, l'esprit de la démonstration.

On donne un groupe G, non trivial, de germes de difféomorphismes en 0, de R dans R, et de classe C^1, admettant une famille finie de générateurs $\{a_1(x), b_1(x), \ldots, a_1^{-1}(x), \ldots\}$ où : $a_1(x) = x + a(x)$ avec $a'(0) = 0, \ldots$.

LEMME.- Il existe un homomorphisme non trivial de G dans R.

Choisissons * $\rho \neq 0$ de sorte que $a(\rho) = \alpha, b(\rho) = \beta, \ldots$ ne soient pas tous nuls (si un tel choix était impossible $a(x), \ldots$ seraient nuls).

Soit $\delta = \max\{|\alpha|, \ldots\}$. Il existe des réels uniques $A, \ldots$ tels que :

$$\alpha = (A + \varepsilon)\delta, \ldots .$$

La loi : $a_1 \to A, \ldots$ engendre un homomorphisme non trivial de G dans R. En effet désignant par $u_1(x)$ le composé $b_1(a_1(x))$, par exemple, alors :

$$u_1(x) = x + u(x) \quad \text{et} \quad u_1(\rho) = \rho + \alpha + b(\rho + \alpha) ;$$

dès lors, la formule des accroissements finis donne :

$$u_1(\rho) = \rho + \alpha + \beta + \alpha b'(\rho + t\alpha) \qquad 0 < t < 1$$

$$= \rho + (A + B + \eta)\delta .$$

C.Q.F.D.

Il n'y a évidemment aucune difficulté à remplacer dans l'énoncé (et la démonstration) R par R^k.

---

* Les lettres grecques désignent des quantités (non standard) infiniment petites.

Addendum: UNE MODIFICATION STANDARD DE LA DEMONSTRATION NON STANDARD DE REEB ET SCHWEITZER

par

W. SCHACHERMAYER (Mexico)

Nous donnons une autre démonstration du lemme de Thurston, qui n'utilise qu'une soussuite mais dont l'idée essentielle est identique à celle de la démonstration plus haut. A nôtre avis il est le mérite de l'analyse non standard, qu'elle rend claire la situation et permet de trouver la très simple démonstration suivante, qui remplace le quotient des quantités infiniment petites par une limite des quotients usuels.

Il convient de changer la notation un peu: Nous notons $\{g_1,\ldots,g_k\}$ une famille finie symétrique de générateurs de G et pour $g \in G$ nous posons $g(x)=\bar{g}(x)+x$. Par hypothèse $(\bar{g})'(0)=0$. En abus de notation nous écrivons aussi g pour un représentant arbitraire du germe g.

Autre preuve du lemme: Par un calcul élémentaire on vérifie la formule suivante $\forall\, g,h \in G$ et $\forall\, x \in \mathbb{R}$:

$$(\overline{g\circ h})\ (x) = \bar{g}(x) + \bar{h}(x) + (\bar{g}(x+\bar{h}\ (x)) - \bar{g}(x)) \qquad (*)$$

Editor's Note. This standard proof, inspired by the nonstandard proof given in the preceding note, illustrates the heuristic value of nonstandard mathematics. The nonstandard proof has also led to short standard proofs by J. Jouanolou and by J. Cantwell.

Choisissons maintenant une suite $\{x_n\}_{n=1}^{\infty}$ dans $\mathbb{R}$ tendant vers 0 telle que pour au moins un $i \in \{1,\ldots k\}$ la suite $\{\bar{g}_i(x_n)\}_{n=1}^{\infty}$ soit différente de zéro pour un nombre infini des $n \in \mathbb{N}$.

Posons $M_n = \max\{|\bar{g}_1(x_n)|,\ldots,|\bar{g}_k(x_n)|\}$. Passant a une sous-suite on peut supposer que $M_n > 0$ pour chaque $n \in \mathbb{N}$ et aussi que pour chaque $i \in \{1,\ldots k\}$ la suite $\{\bar{g}_i(x_n)/M_n\}_{n=1}^{\infty}$ converge vers un élément de $[-1, +1]$, disons vers b_i. Pour un certain $i_o \in \{1,\ldots,k\}$ nous avons alors $|b_{i_o}| = 1$.

Si g est un élément arbitraire de G tel que $\{\bar{g}(x_n)/M_n\}_{n=1}^{\infty}$ converge (disons vers b), alors pour chaque générateur g_i la suite $\{(\overline{g_i \circ g})(x_n)/M_n\}_{n=1}^{\infty}$ converge vers $b+b_i$.

En effet en vue de la formule (*) il suffit de vérifier

$$\lim_{n\to\infty} M_n^{-1}(\bar{g}_i(x_n+\bar{g}(x_n)) - \bar{g}_i(x_n)) = 0 \qquad (**)$$

Utilisant le théorème de valeur moyenne nous avons

$$\lim_{n\to\infty} M_n^{-1}(\bar{g}_i(x_n+\bar{g}(x_n)) - \bar{g}_i(x_n)) =$$

$$\lim_{n\to\infty} M_n^{-1}\,\bar{g}(x_n)\,((\bar{g}_i)'(s_n))$$

où s_n note un point entre $x_n+\bar{g}(x_n)$ et x_n. Le premier terme est une suite convergente et le deuxième tend vers zéro, parce que s_n tend vers zéro et par hypothèse $(\bar{g}_i)'$ est continu et zéro en zéro, ce qui démontre (**).

Si l'on pose alors

$$H: \quad G \to \mathbb{R}$$

$$g \mapsto \lim_{n\to\infty} \bar{g}(x_n)/M_n \quad ,$$

H est bien défini et c'est un homomorphisme non-trivial.

Remarque: Evidemment cet argument se généralise aussi immédiatement à $\mathbb{R}^k$.

CROISSANCE DES FEUILLETAGES PRESQUE SANS HOLONOMIE

Gilbert HECTOR

Le type de croissance des feuilles d'un feuilletage de codimension 1 est évidemment lié à la nature de l'holonomie de ce feuilletage. Le présent travail se propose d'expliciter ce rapport dans le cas particulier des feuilletages presque sans holonomie. Il comprend six parties :

I) Introduction - Résultats principaux.

II) Un lemme de trivialisation.

III) Structure des feuilletages presque sans holonomie.

IV) Croissance des feuilletages presque sans holonomie.

V) Feuilletages à croissance d'ordre inférieur ou égal à 1.

VI) Exemples de croissance.

Ce travail qui a trouvé son origine dans le problème n° 7 de [12] est le résultat d'un séjour à l'I.M.P.A. de Rio de Janeiro .... L'attention et les encouragements des participants au "Seminario de Folheaçoes" m'ont puissamment stimulé dans cette étude. La liste de tous leurs noms serait longue... mieux vaut aller faire leur connaissance sur place. Je remercie également J. Plante pour les remarques qu'il a bien voulu me faire.

I - INTRODUCTION - RESULTATS PRINCIPAUX.-

L'expression "croissance polynomiale" sera utilisée ici dans un sens un peu plus restrictif que d'ordinaire ; aussi, pour fixer les idées, nous allons rappeler très brièvement les définitions usuelles relatives à la croissance des fonctions [resp. des feuilletages] .

a) Soient f et g deux éléments de $C(\mathbb{R}^+)$, l'ensemble des applications croissantes de $\mathbb{R}^+$ dans lui-même (de variable x).

i) On dit que f est dominée par g s'il existe trois constantes positives α,β,δ telles que $f(x) \leq \alpha g(\beta x + \delta)$ pour tout $x \in \mathbb{R}^+$.

ii) On note croiss(f) la classe de f pour la relation d'équivalence associée au préordre défini en (i) et deux éléments équivalents sont dits "avoir même type de croissance".

iii) On munit l'ensemble des types de croissance de la relation d'ordre induite, que l'on notera $\leq$, et pour $n \in \mathbb{N}$, on dit que f est à croissance d'ordre inférieur ou égal à n [resp. à croissance polynomiale de degré n] si on a $\text{croiss}(f) \leq \text{croiss}(x^n)$ [resp. $\text{croiss}(f) = \text{croiss}(x^n)$]. (Dans le second cas, on écrira aussi $\deg(f) = n$) .

iv) Enfin f est à croissance exponentielle si on a : $\text{croiss}(f) \geq \text{croiss}(e^x)$ et à croissance non-exponentielle si non.

b) Si F est une variété riemannienne orientable de classe C^1 et a un point de F, on note $B_a(r)$ pour $r \in \mathbb{R}^+$ la boule fermée de centre a et rayon r dans F et g_a la fonction croissance (géométrique) de F en a, définie par $g_a(r) = \text{vol}\, B_a(r)$. Cette fonction dépend du point a et de la métrique de F. Par contre, si F est une feuille d'un feuilletage $\mathcal{F}$ sur une variété compacte M munie de la métrique induite, le type de croissance de g_a, que l'on appelle encore type de croissance de F et que l'on note croiss(F) est parfaitement défini.

c) Enfin, un feuilletage $\mathcal{F}$ sur M est dit à croissance d'ordre inférieur ou égal à n [resp. à croissance polynomiale de degré n] si toutes ses feuilles sont à croissance d'ordre inférieur ou égal à n [resp. à croissance polynomiale telle que $n = \sup_{F \in \mathcal{F}} \deg(F)$]. (Dans le second cas, on écrira aussi $n = \deg(\mathcal{F})$). Et bien sûr, $\mathcal{F}$ est dit à croissance non-exponentielle s'il en est ainsi pour toutes ses feuilles.

On sait que si $\mathcal{F}$ est un feuilletage de codimension 1, de classe C^2 sur une variété compacte M et si $\mathcal{F}$ est à croissance non-exponentielle, toutes les feuilles de $\mathcal{F}$ sont partout denses et sans holonomie si aucune d'entre elles n'est compacte. Réciproquement, si $\mathcal{F}$ est défini par une action localement libre de classe C^2 de $\mathbb{R}^{n-1}$ sur M, $\mathcal{F}$ est à croissance non-exponentielle et vérifie les conditions suivantes :

(p_1) $\mathcal{F}$ est presque sans holonomie (en abrégé p.s.h.) , i.e. , l'holonomie de toute feuille non compacte est triviale ([10]).

(p_2) $\mathcal{F}$ ne possède pas de feuille exceptionnelle.

Nous nous proposons essentiellement ici de montrer que (p_1) implique (p_2) et que la croissance des feuilletages presque sans holonomie est non-exponentielle (et même polynomiale) mais que la réciproque n'est pas vraie.

1 - Structure des feuilletages presque sans holonomie.

Dans la suite, on désignera par (M,F,N) ou simplement (M,F) ou F, un feuilletage F de codimension 1, de classe C^r, $r \geq 0$, transversalement orienté par un champ de vecteurs N, sur une variété riemannienne compacte M de dimension $(n+1)$, tangent au bord si $\partial M \neq \emptyset$.

La première définition utilisée est un peu différente de la définition correspondante de [4].

Définition 1.- *Un feuilletage p.s.h. (M,F,N) est un modèle (de feuilletages p.s.h.) de type 1 ou 2 si la condition correspondante est satisfaite :*

1) le feuilletage $\mathring{F}$ induit par F dans l'intérieur $\mathring{M}$ de M est sans holonomie ;

2) M est un produit de la forme $L \times [0,1]$ (où L est une variété de dimension n) et N est tangent au facteur $[0,1]$.

On ramène la description des feuilletages p.s.h. à celle des modèles

Théorème 1.- *Soit (M,F,N) un feuilletage p.s.h. de classe C^r, $r \geq 0$. Il existe une famille finie (M_i,F_i,N_i), $i \in \{1,\ldots,p\}$ de modèles et une surjection $\emptyset$ de classe C^r de la somme disjointe $M_* = \coprod_{i=1}^{p} M_i$, munie des feuilletages "somme disjointe" F_* et N_*, sur M tels que :*

i) pour tout i, $\emptyset(M_i)$ est une sous-variété compacte de M et la *restriction de* ψ *à* $\overset{\circ}{M}_i$ *et à chaque composante connexe de* ∂M_i *est injective ;*

ii) $\psi^*(F) = F_*$ *et* $\psi^*(N) = N_*$.

Un modèle de type 2 sur $M = L\times[0,1]$ est complètement déterminé par une représentation du groupe fondamental de L dans le groupe des homéomorphismes croissants de $[0,1]$ (cf [2]) dont l'image G est un groupe abélien (cf. lemme 5). Un exemple de modèle de type 1 qui n'est pas en même temps de type 2, est le feuilletage de Reeb sur $D^2 \times S^1$. De façon générale, on va voir que si (M,F) est un modèle de type 1, de classe C^2, le feuilletage $\overset{\circ}{F}$ est une fibration de $\overset{\circ}{M}$ sur S^1 ou bien toutes ses feuilles sont partout denses (cf. lemme 7). La notion de métrique "bundle-like" introduite par B.L. Reinhart (cf.[19]) permet d'obtenir une description plus précise dans ce dernier cas.

<u>Définition</u> 2.- *Une métrique riemannienne* R *sur* (M,F,N) *est dite* *<u>"bundle-like"</u>* *si :*

i) N *est orthogonal à* F *pour* R *;*

ii) la métrique induite par R *sur la somme disjointe des courbes intégrales de* N *est invariante par le pseudo-groupe d'holonomie de* F *(cf. introduction du paragraphe III).*

<u>Théorème</u> 2.- *Si* (M,F) *est un modèle de type 1, de classe* C^2, *il existe une structure différentiable* S_F *et une métrique riemannienne "bundle-like"* R_F *de classe* C^2 *sur* $\overset{\circ}{M}$ *telles que :*

i) les structures induites par S_F *et* R_F *sur chaque feuille de* $\overset{\circ}{F}$ *sont égales aux structures initiales ;*

ii) $\overset{\circ}{F}$ *est défini par une forme fermée dans la structure* S_F *;*

iii) $\overset{\circ}{M}$ *est complète pour la métrique* R_F.

Ce théorème, qui se réduit au théorème 6 de [20] lorsque $\partial M = \emptyset$, se démontre de façon analogue à celui-ci. Et comme l'adhérence de toute composante connexe de la réunion des feuilles non compactes d'un feuilletage p.s.h. est un modèle de type 1, on en déduit immédiatement, compte tenu du lemme 5, le résultat suivant :

<u>Théorème</u> 3.- *Si* (M,F) *est un feuilletage p.s.h. de classe* C^2, *on a :*

(p_2) F *ne possède pas de feuille exceptionnelle ;*

(p_3) *le groupe d'holonomie de toute feuille de* F *est abélien.*

<u>Remarque</u> *1* . Pas plus que (p_2) . la condition (p_3) n'implique (p_1) . En effet, dans [5], on construit un feuilletage analytique de $V_2 \times S^1$ (où V_2 désigne la surface compacte de genre 2) sans feuille compacte et sans feuille localement dense. Tous les groupes d'holonomie de ce feuilletage sont isomorphes à $\{0\}$ ou $\mathbb{Z}$ d'après le théorème 10 de [6] et ce feuilletage possède au moins une feuille exceptionnelle dont l'holonomie est isomorphe à $\mathbb{Z}$ (cf. [20]).

La démonstration du théorème 2 nécessitera quelques préliminaires de type "descriptif" ; le plus important démontré pour la première fois dans [3] s'énonce comme suit :

<u>Lemme de trivialisation</u> (*).- *Soit* J *un arc de courbe intégrale de* N, *homéomorphe à* $(-1,+1)$ *tel que la relation* ρ_J *induite par* F *sur* J *soit triviale.*

(*) Remarquons que P.R. Dippolito donne, de façon indépendante, une version un peu plus sophistiquée du lemme de trivialisation dans le théorème 1 de [1].

La restriction de F *à l'ouvert* $\mathrm{sat}_F(J)$, *saturé de* J *par* F, *est une fibration triviale de base* J.

2 - Croissance des feuilletages presque sans holonomie.

Venons-en maintenant aux résultats concernant la croissance.

Théorème 4.- *Tout modèle* (M,F) *de classe* C^2 *est à croissance polynomiale et si* $b_1(M)$ *est le premier nombre de Betti de* M, *on a :*

$$\deg(F) \leq b_1(M) \qquad si \quad \partial M \neq \emptyset \ ;$$

$$\deg(F) \leq b_1(M)-1 \qquad si \quad \partial M = \emptyset \ .$$

De plus, si (M,F) *est de type 1 et si* $\partial M \neq \emptyset$, $\overset{\circ}{F}$ *est une fibration de* $\overset{\circ}{M}$ *sur* $\mathbb{S}^1$ *si et seulement si* $\deg(F) = 1$.

On pourrait déduire directement du théorème 2 et du théorème 6.3 de [17] que F est à croissance d'ordre fini (i.e. à croissance polynomiale au sens de Plante). La partie difficile du théorème 4 consiste à montrer que F est à croissance polynomiale (au sens que nous avons adopté ici).

Le théorème 4 montre par exemple que le feuilletage de Reeb sur $\mathbb{S}^3$ est à croissance linéaire.

Théorème 5.- *Tout feuilletage* (M,F) *presque sans holonomie de classe* C^2 *est à croissance polynomiale*. (*)

De plus, si F *possède une feuille compacte,* $\deg(F) = 1$ *si et seulement si toutes les feuilles de* F *sont propres.*

Remarque 2.- Lorsque F ne possède pas de cycle évanouissant, il doit être possible de trouver une majoration de $\deg(F)$ en fonction de $b_1(M)$.

(*) Pour la croissance des groupes, on se rapportera au début du paragraphe IV.

3 - *Autres propriétés de croissance.*

Le théorème bien connu de J. Plante sur la croissance des feuilles d'un minimal exceptionnel [resp. des feuilles minimales denses à holonomie non triviale] (cf. [15], [17]) s'étend au cas des feuilles-ressort, qui sont les feuilles qui "spiralent" sur elles-mêmes (cf. [7], voir aussi paragraphe V).

Théorème 6.- *Toute feuille - ressort d'un feuilletage F de classe C^r, $r \geq 0$, est à croissance exponentielle.*

Et le théorème 5 admet la réciproque partielle suivante :

Théorème 7.- *Si F est un feuilletage de classe C^2 dont la croissance est d'ordre inférieur ou égal à 1, on a :*

i) F est à croissance polynomiale de degré 0 ou 1 ;

ii) F est presque sans holonomie ;

iii) toutes les feuilles de F sont propres si l'holonomie de F n'est pas triviale.

De fait, cette réciproque est la meilleure possible ; car si V_2 désigne la surface compacte de genre 2, on a :

Théorème 8.- *Pour tout $n \in \mathbb{N}^*$, il existe un feuilletage F_n de classe C^∞ sur $V_2 \times S^1$, transverse au facteur S^1 tel que :*

i) toutes les feuilles de F_n sont propres ;

ii) pour tout entier positif $p < n$, il existe exactement une feuille $F(p) \in F_n$ de degré p à holonomie non triviale et toutes les autres feuilles de F_n sont difféomorphes, sans holonomie et de degré n. Bref F_n est à croissance polynomiale de degré n.

Tous les exemples annoncés au théorème 8 sont construits à l'aide de groupes de difféomorphismes de S^1 à croissance exponentielle. Le feuilletage F_2 est quadratique mais possède une feuille non compacte à holonomie non triviale . Un dernier résultat va nous donner quelques indications sur le comportement possible des fonctions croissances pour des feuilletages plus complexes.

Théorème 9.- *Il existe un feuilletage* F_b [*resp.* F_c] *de classe* C^∞ *de* $V_2 \times S^1$, *transverse à* S^1 *tel que :*

i) pour tout $p \in N$, F_b [*resp.* F_c] *possède exactement une* [*resp.* 2^p] *feuilles* $F(p)$ *à holonomie non triviale et à croissance de degré* p ;

ii) la réunion $\bigcup_{p=0}^{\infty} F(p)$ *est dense dans* $V_2 \times S^1$; *toutes les feuilles du complémentaire sont difféomorphes, partout denses* [*resp. exceptionnelles*] *et l'une d'entre elles (au moins) est à croissance exponentielle.*

En guise de conclusion, nous voudrions poser les questions suivantes :

Problème n° 1.- Une feuille à croissance d'ordre fini est-elle polynomiale ?

Problème n° 2.- Existe-t-il des feuilles à croissance non-exponentielle et d'ordre infini ?

Problème n° 3.- Peut-on caractériser géométriquement (i.e. par exemple par des propriétés d'holonomie) les feuilletages à croissance polynomiale ? non-exponentielle ?

II - *LEMME DE TRIVIALISATION*.-

Un ouvert ω de (M,F,N) est dit bidistingué si ω est homéomorphe à $\mathbb{R}^n$ et si :

i) ω est distingué (au sens habituel) à la fois pour les feuilletages F et N ;

ii) toute F-plaque de ω (i.e. toute plaque de ω au sens du feuilletage F) coupe toute N-plaque exactement en un point.

La famille des ouverts bidistingués constitue évidemment une base de la topologie de M.

On définit au § 1 des opérateurs de "relèvement des chemins" : les projecteurs et quasi-projecteurs. Ils seront utilisés au § 2 pour démontrer le lemme de trivialisation, qui sera utilisé lui dans III.

1 - *Projecteurs et quasi-projecteurs* (cf [21], [3]).

Pour $m \in M$, on note F_m [resp. N_m] la feuille de F [resp. la courbe intégrale de N] passant par m.

On désigne par $A = [a_o, a_1]$ et $B = [b_o, b_1]$ deux intervalles fermés bornés de $\mathbb{R}$, par AB leur produit cartésien et par AB^* le produit AB privé du sommet (a_1,b_1). Un couple (σ,τ) de chemins dans M paramétrés respectivement par A et B forme un cornet-régulier de sommet m si

(c_1) $\sigma(a_o) = \tau(b_o) = m$;

(c_2) $\sigma(A) \subset F_m$ et $\tau(B) \subset N_m$;

(c_3) τ est une isométrie locale.

Les points $\sigma(a_1)$ et $\tau(b_1)$ sont les extrémités du cornet.

Définition 3 (cf. [21]).- *Soit* (σ,τ) *un cornet régulier. Une application* $P : AB \to M$ *est un <u>projecteur engendré par</u>* (σ,τ) *si on a*

i) $P(s,b_o) = \sigma(s)$ *pour tout* $s \in A$ *et*

$P(a_o,t) = \tau(t)$ *pour tout* $t \in B$;

ii) $P(s,t) \in F_{\tau(t)} \cap N_{\sigma(s)}$ *pour tout* $(s,t) \in AB$.

On définit de façon analogue un <u>quasi-projecteur engendré par</u> (σ,τ) *en remplaçant* AB *par* AB^* *dans la définition précédente.*

En fait, il est facile de voir qu'un projecteur [resp. quasi-projecteur] engendré par (σ,τ) est unique (s'il existe). De plus il est de classe C^r si F, N, σ et τ sont de classe C^r. Enfin si $P : AB \to M$ est un projecteur, la restriction Q de P à AB^* est un quasi-projecteur : on dit que Q est <u>prolongeable</u>. Il est clair qu'un quasi-projecteur quelconque n'est pas prolongeable en général. Toutefois si on note $Q_1 : [b_o,b_1) \to M$ la restriction de Q à $\{a_1\} \times [b_o,b_1)$, on peut donner quelques conditions simples de prolongement des quasi-projecteurs :

Lemme 1.- Un quasi-projecteur Q *est prolongeable si et seulement si* Q_1 *se prolonge à* $[b_o,b_1]$ *(en tant qu'application continue).*

Démonstration.- Supposons que Q_1 se prolonge en une application $\bar{Q}_1$ et soient $m = \bar{Q}_1(b_1)$ et ω un voisinage ouvert bidistingué de m. Si Q est engendré par le cornet (σ,τ), on peut supposer, sans restriction de généralité, que σ et $\bar{Q}_1$ sont à valeurs dans ω. Mais alors il en est de même pour τ donc pour Q. Par suite, d'après la définition des ouverts bidistingués, Q est prolongeable. D'où le lemme.

Lemme 2.- Soit Q *un quasi-projecteur non prolongeable. Si* Q *est engendré par le cornet* (σ,τ), *il existe (au moins) deux éléments* $t_o < t_1$

appartenant à $[b_o,b_1)$ *tels que* $\tau(t_o)$ *et* $\tau(t_1)$ *appartiennent à la même feuille de* F.

<u>*Démonstration*</u>.- D'après le lemme 1, si Q n'est pas prolongeable, il en est de même pour Q_1 qui est une application monotone de $[b_o,b_1)$ dans $N = N_{\sigma(a_1)}$. Or N est complète, puisque M est compacte et donc Q_1 n'est pas uniformément continue. Deux cas sont alors à distinguer :

1) N est compacte donc homéomorphe à $\mathbb{S}^1$ et Q_1 décrit N une infinité de fois ; en particulier il existe $t_o < t_1$ tels que $Q_1(t_o) = Q_1(t_1)$.

2) N n'est pas compacte et $Q_1((b_o,b_1))$ est l'une des composantes connexes de $N - \{\sigma(a_1)\}$; par compacité de M, il existe un ouvert bidistingué ω tel que $Q_1((b_o,b_1))$ rencontre ω suivant au moins deux N-plaques et il existe $t_o < t_1$ tels que $Q_1(t_o)$ et $Q_1(t_1)$ appartiennent à la même F-plaque de ω. Donc dans les deux cas le lemme est démontré.

<u>*Lemme*</u> 3.- *Soit* W *un ouvert saturé pour* F *de* (M,F,N). *Les deux conditions suivantes sont équivalentes :*

i) tout quasi-projecteur Q *à valeurs dans* W *est prolongeable ;*

ii) tout cornet régulier (σ,τ) *à valeurs dans* W *engendre un projecteur* P.

<u>*Démonstration*</u>.- Il suffit bien sûr de montrer que (i) implique (ii). Pour cela, soit (σ,τ) un cornet régulier paramétré par (A,B). On ordonne le produit AB par la relation d'ordre lexicographique :

$$(a,b) \leq (a',b') \text{ si } a < a' \text{ ou } a = a' \text{ et } b \leq b' .$$

De plus pour $a \in A$ [resp. $b \in B$], on note σ_a [resp. τ_b] la restriction de σ à $[a_o,a]$ [resp. de τ à $[b_o,b]$] . Pour tout $(a,b) \in AB$, le couple

(σ_a,τ_b) est évidemment un cornet régulier. De plus l'ensemble $P_{(\sigma,\tau)} \subset AB$ des couples (a,b) tels que (σ_a,τ_b) engendre un projecteur est un intervalle ouvert non vide de AB (au sens de la topologie de l'ordre). Enfin si (α,β) est la borne supérieure de P, le cornet correspondant $(\sigma_\alpha,\tau_\beta)$ engendre un quasi-projecteur Q.

Par suite si W vérifie (i) et si (σ,τ) est à valeurs dans W, on a $P_{(\sigma,\tau)} = AB$ donc (σ,τ) engendre un projecteur et W vérifie (ii).

2 - Lemme de trivialisation.

Soit L une variété de dimension n. On note H_L [resp. V_L] le feuilletage de $L \times (-1,+1)$ dont les feuilles sont les variétés de la forme $L \times \{t\}$, $t \in (-1,+1)$ [resp. $\{x\} \times (-1,+1)$, $x \in L$].

Définition 4.-

a) Un ouvert W *de* (M,F,H) *, saturé pour* F*, est dit* *feuilleté en produit par* (F,N) *s'il existe une variété* L *et un homéomorphisme* ϕ *de* $L \times (-1,+1)$ *sur* W *tels que* $\phi^*(F) = H_L$ *et* $\phi^*(N) = V_L$.

b) On dira en outre qu'un voisinage à droite [resp. à gauche] $\hat{W}$ *d'une feuille* F *de* F *est un* *voisinage-collier* *si* $W = \hat{W} - F$ *est feuilleté en produit par* (F,N).

Une feuille qui possède un voisinage-collier à droite [resp. à gauche] est évidemment propre à droite [resp. à gauche] ; elle peut néanmoins être exceptionnelle ; nous dirons qu'elle est semi-propre

Nous pouvons maintenant donner un énoncé plus précis du lemme de trivialisation ainsi qu'une version améliorée du "main-lemma" de [21], la démonstration de ce dernier valable pour les feuilles propres s'étendant sans problème au cas des feuilles semi-propres.

<u>Lemme de trivialisation</u>.- *Soit* J *un arc de courbe intégrale de* N*, homéomorphe à* $(-1,+1)$ *tel que la relation* ρ_J *induite par* F *sur* J *soit triviale.*

Le saturé W *de* J *par* F *est un ouvert feuilleté en produit par* (F,N).

<u>Démonstration</u>.- Comme ρ_J est triviale, le feuilletage F_W induit par F dans W est sans holonomie.

Soient alors $m_o \in J$ et $L = F_{m_o}$. Pour tout $(x,v) \in L \times J$, il existe un cornet régulier (σ,τ) de sommet m_o et d'extrémités (x,v). D'après les lemmes 2 et 3, ce cornet engendre un projecteur P à valeurs dans W et F_W étant sans holonomie, le quatrième sommet de P ne dépend pas du choix de (σ,τ) ; on le note $\psi(x,v)$. L'application $\psi : L \times J \to W$ ainsi définie est de façon évidente un homéomorphisme local tel que $\psi^*(F) = H_L$ et $\psi^*(N) = V_L$. Il nous reste donc à voir que ψ est bijective.

Comme ρ_J est triviale, si (x,v) et (x',v') sont deux points de $L \times J$ tels que $\psi(x,v) = \psi(x',v')$, on a $v = v'$ et d'après la nullité de l'holonomie de F_W on a également $x = x'$ et donc ψ est injective. En outre, pour tout $m \in W$, il existe un cornet régulier $(\tilde{\sigma},\tilde{\tau})$ de sommet $p \in J$ et d'extrémités (m,m_o). Comme précédemment $(\tilde{\sigma},\tilde{\tau})$ engendre un projecteur $\tilde{P}$ paramétré par un produit AB. Alors si on appelle σ la restriction de $\tilde{P}$ à $A \times \{b_1\}$ et si $\tau = (\tilde{\tau})^{-1}$, le point m est le quatrième sommet du projecteur P engendré par (σ,τ). Bref ψ est surjective, d'où le lemme.

<u>Proposition 1</u>.- *Pour toute feuille* F *propre à droite [à gauche] de* (M,F,N)*, on a l'une des deux situations suivantes :*

i) F *possède un voisinage-collier à droite [resp. à gauche]*

ii) F *est adhérente à droite [resp. à gauche] à la réunion des feuilles à holonomie non triviale.*

III - STRUCTURE DES FEUILLETAGES PRESQUE SANS HOLONOMIE.-

Soit $M(N)$ la somme disjointe des courbes intégrales de N. Tout chemin contenu dans une feuille de F définit, par relèvement dans les feuilles voisines le long de N, un homéomorphisme local de $M(N)$. L'ensemble Γ de tous ces homéomorphismes locaux est le pseudo-groupe d'holonomie de F. Il est de classe C^r si F et N sont de classe C^r. Une métrique R sur M est dite invariante par Γ si la métrique induite sur $M(N)$ est invariante par Γ.

Le pseudo-groupe Γ est énorme, mais on peut le réduire à un pseudo-groupe de type fini (cf. [17]). A cet effet, on utilise un recouvrement ouvert bidistingué régulier i.e. un recouvrement fini $\Omega = \{\omega_1, \dots, \omega_p\}$ de M tel que :

(Ω_1) pour tout i, ω_i est ouvert bidistingué et il existe un ouvert bidistingué ω_i' tel que $\overline{\omega_i} \subset \omega_i'$;

(Ω_2) si $\omega_i \cap \omega_j \neq \emptyset$, toute F-plaque [resp. N-plaque] de ω_i rencontre au plus une F-plaque [resp. N-plaque] de ω_j .

Alors si pour tout i, on suppose choisie une N-plaque X_i de ω_i (que l'on appellera l'axe de ω_i), les conditions de régularité permettent de définir un homéomorphisme local $h_{ji} : X_i \to X_j$ si $\omega_i \cap \omega_j \neq \emptyset$. Et l'ensemble fini $\Sigma = \{h_{ji}\}$ engendre un pseudo-groupe Γ_X d'homéomorphismes locaux de la somme disjointe X des X_i que l'on appellera le pseudo-groupe d'holonomie de F réduit à X (X s'appellera également l'axe de Ω).

Remarquons que si on note $\Gamma_{(X,x)}$ le sous-pseudo-groupe d'isotropie de Γ_X en un point $x \in X$, le groupe d'holonomie de la feuille F_x au point x (au sens usuel) n'est rien d'autre que le groupe des germes en x des éléments de $\Gamma_{(X,x)}$.

Pour finir, tout élément de Γ_X étant un mot sur Σ, on peut définir la longueur d'un élément de Γ_X (par rapport à Σ) ; ce qui induit sur toute trajectoire T de Γ_X une distance δ en posant :

$$\delta(x,y) = \inf_{g\in\Gamma_X} \{\mathrm{long}(g) \mid g(x) = y\}.$$

Pour $x \in X$, on appellera raccourci en x tout élément $g \in \Gamma_X$ réalisant la distance de x à $g(x)$.

Au §1, on commence par étendre le champ d'application d'un théorème bien connu de R. Sacksteder. Ensuite on démontre les théorèmes 1 au §2 et 2 au §3.

1 - *A propos du théorème du point fixe de Sacksteder* (cf. [20]).

Si F est une feuille semi-propre de $\mathcal{F}$, on peut choisir un recouvrement bidistingué régulier $\Omega = \{\omega_1,\dots,\omega_p\}$ de M de manière à vérifier la condition supplémentaire :

$$(\Omega_F) \quad \bar{X}_i \cap \bar{F} = X_i \cap \bar{F} \quad \text{pour tout} \quad i \in \{1,\dots,p\} .$$

Alors bien que $\bar{F}$ ne soit pas minimale en général, on peut extraire de la démonstration du théorème 1 de [20], le résultat partiel suivant :

Proposition 2.- *Si* F *est une feuille semi-propre d'un feuilletage* $\mathcal{F}$ *de classe* C^2, *si le recouvrement* Ω *vérifie* (Ω_F) *et si* a *est un point fixé de* $F \cap X$, *il existe un voisinage* V *de* a *dans* X *et une suite décroissante* $\{k_n\}_{n\in\mathbb{N}}$ *convergeant vers* 0 *tels que :*

i) tout raccourci g *en* a *appartenant à* Γ_X *est défini sur* V *;*

ii) si g *est de longueur* n, *on a* $g'(v) \leq k_n$ *pour tout* $v \in V$.

En effet, dans la démonstration de ce résultat, R. Sacksteder utilise seulement la propriété (Ω_F) , le fait que X est de mesure finie et,

F étant de classe C^2, l'existence de deux constantes positives λ et μ telles que :

$$\mu^{-1} \leq h'(x) \leq \mu \quad \text{et} \quad |h''(x)| \leq \lambda \mu^{-1}$$

pour tout $h \in \Sigma$ et tout x appartenant au domaine D_h de h. Or cette dernière condition découle de la régularité du recouvrement Ω.

Bien plus le théorème 1 de [20] lui-même s'étend alors sans difficulté au cas des feuilles semi-propres, sous la forme suivante :

Théorème de Sacksteder.- *Si* F *est une feuille semi-propre exceptionnelle d'un feuilletage* F *de classe* C^2*, sur une variété compacte* M*, il existe pour tout voisinage ouvert saturé* U *de* F*, un point* $x_U \in X \cap \bar{F} \cap U$ *et un élément* $g_U \in \Gamma_X$ *tels que :*

$$g_U(x_U) = x_U \quad \text{et} \quad g'_U(x_U) < 1 .$$

2 - Réduction des feuilletages presque sans holonomie aux modèles.

D'après le théorème 3.2 de [2], la réunion A des feuilles compactes d'un feuilletage (M,F,N) est un fermé de M. De plus, d'après le théorème bien connu de Haefliger, on peut supposer, quitte à couper M le long d'un nombre fini de feuilles compactes, qu'aucune feuille compacte n'est coupée par une transversale fermée. Enfin, si $\{W_n\}_{n\in\mathbb{N}}$ est la suite des composantes connexes de M-A et si $\bar{F}_n$ est le feuilletage induit par F sur $\bar{W}_n$, $(\bar{W}_n,\bar{F}_n)$ est un modèle de type 1 pour tout n si F est presque sans holonomie.

Lemme 4.- *Pour presque tout* n*,* (W_n,F_n) *est un modèle de type 2.*

Démonstration.- En effet, dans le cas contraire, il existe une sous-suite $\{\bar{W}_{n_j}\}$ de $\{\bar{W}_n\}$ telle que $(\bar{W}_{n_j}, F_{n_j})$ n'est pas modèle de type 2. Pour tout n_j, choisissons une feuille compacte $F_{n_j} \subset \bar{W}_{n_j}$; quitte à extraire une nouvelle suite partielle, nous pouvons supposer que la suite $\{F_{n_j}\}$ possède une feuille limite compacte F_*. Alors pour presque tout n_j, $\bar{W}_{n_j}$ est contenue dans un voisinage tubulaire de F_*, donc $\bar{W}_{n_j}$ est un modèle de type 2 ; ce qui amène une contradiction.

Démonstration du théorème 1.- Soit (M,F) un feuilletage p.s.h.

a) Si $A = \emptyset$, (M,F) est un modèle de type 1.

b) Si $A \neq \emptyset$ et si $(\bar{W}_n, \bar{F}_n)$ est un modèle de type 2 pour tout n, toute feuille compacte de F possède un voisinage fermé saturé qui est un modèle de type 2. Il est alors facile de recouvrir M par une famille finie $\{V_1, V_2, \ldots, V_s\}$ de telles sous-variétés et la famille $\{V_1, \overline{V_2 - V_1}, \ldots, \overline{V_s - (\bigcup_{i=1}^{s-1} V_i)}\}$ définit une décomposition de (M,F) en modèles.

c) Dans le cas général enfin, si $\{W_1, \ldots, W_p\}$ est la famille finie des composantes de $M-A$ qui définissent des modèles de type 1 qui ne sont pas de type 2, on applique (b) à la variété $M - (\bigcup_{i=1}^{p} W_i)$. D'où le théorème.

3 - *Structure des modèles*.

Une première propriété des feuilletages p.s.h. est bien connue :

Lemme 5.- *Le groupe d'holonomie de toute feuille* F *d'un feuilletage p.s.h. est abélien.*

Démonstration.- Il nous suffit de montrer que le groupe d'holonomie à droite [resp. à gauche] d'une feuille compacte F est abélien.

Or si F est isolée à droite dans l'ensemble des feuilles compactes, le groupe d'holonomie à droite G^+ de F peut être réalisé par un pseudo-groupe d'homéomorphismes locaux de $[0,+\infty)$ définis au voisinage de 0 et qui sont soit des contradictions soit des dilatations (ou encore l'identité). Par suite G^+ est totalement ordonné, archimédien donc abélien.

Si F n'est pas isolée à droite, G^+ peut être réalisé par un groupe d'homéomorphismes de $[0,1]$ qui ont tous le même ensemble B de points fixes et qui commutent sur les composantes connexes du complémentaire de B. Bref, dans ce cas également G^+ est abélien. D'où le lemme.

Une construction classique montre que si (M,F) est un modèle de type 1, toute feuille du feuilletage $\overset{\circ}{F}$ induit par F dans $\overset{\circ}{M}$ est coupée par au moins une transversale fermée θ à F. Bien plus, on a :

__Lemme__ *6.- Si* (M,F) *est un modèle de type 1, le saturé* Θ *par* F *de toute transversale fermée* θ *à* F *est égal à* $\overset{\circ}{M}$.

__*Démonstration*__.- L'ensemble Θ est évidemment un ouvert de $\overset{\circ}{M}$. Soit $\hat{\Theta}$ l'ensemble des points x de M qui possèdent un voisinage simplement connexe V_x dans N_x tel que l'une (au moins) des composantes connexes de $V_x - \{x\}$ soit contenue dans Θ. On a bien sûr $\hat{\Theta} \subset \bar{\Theta}$.

En outre, soit $x \in \hat{\Theta} - \Theta$; la feuille F_x est semi-propre (mettons propre à droite) et ne possède pas de collier à droite, car sinon, d'après le lemme de trivialisation, la transversale θ rencontrerait F_x et on aurait $x \in \Theta$. Par suite, d'après la proposition 1, F_x est adhérente à droite à la réunion des feuilles à holonomie non triviale et comme $\overset{\circ}{F}$ est sans holonomie, ceci implique que l'holonomie de F_x n'est pas triviale, autrement dit F_x est compacte. On en déduit que $\hat{\Theta} = \bar{\Theta} = M$. D'où le lemme.

Quitte à modifier N, nous supposerons désormais que N possède une courbe intégrale fermée θ dans $\overset{\circ}{M}$ et on note ρ_θ la relation d'équivalence induite par F sur θ.

__Lemme__ 7.- *Si* (M,F) *est un modèle de type 1 de classe* C^2*, on a, quitte à modifier légèrement* θ*, une des deux situations suivantes :*

i) ρ_θ *est triviale ;*

ii) toutes les trajectoires de ρ_θ *sont partout denses.*

__Démonstration.__- La relation ρ_θ possède au moins un ensemble minimal M_θ qui est de l'un des trois types bien connus :

i) une trajectoire finie ;

ii) la transversale θ ;

iii) un ensemble minimal exceptionnel .

Mais en fait, le cas (iii) est exclu par le théorème de Sacksteder et dans le cas (i) toutes les trajectoires de ρ_θ sont finies de même ordre d'après le lemme 6 ; on peut donc modifier θ de telle manière que ρ_θ soit triviale. D'où le lemme.

__Démonstration du théorème__ 2.- Soit (M,F) un modèle de type 1 de classe C^2.

a) Compte tenu du lemme 7, on établit l'existence des structures S_F et R_F exactement comme dans la démonstration du théorème 6 de [20]. Par construction, S_F et R_F induisent les structures initiales dans chaque feuille F de $\overset{\circ}{F}$ et $\overset{\circ}{F}$ est défini par une forme fermée dans $(\overset{\circ}{M},S_F)$. Il ne nous reste donc plus qu'à montrer que $(\overset{\circ}{M},R_F)$ est complète.

b) Pour cela remarquons que si V est la réunion d'une famille convenable de voisinages tubulaires, deux à deux disjoints, des différentes composantes connexes de ∂M, on peut supposer que l'on a $R \leq R_F$ en restriction à V , (où R désigne la métrique initiale de M). Par compacité de $\overset{\circ}{M}-V$, il existe alors une constante positive λ telle que :

$$(\Lambda) \quad : \quad \lambda R \leq R_F \quad \text{sur} \quad \overset{\circ}{M} \; .$$

On en déduit immédiatement que toute suite de Cauchy dans $(\overset{\circ}{M},R_F)$ est de Cauchy dans $(\overset{\circ}{M},R)$ donc convergente vers un point qui appartient à $\overset{\circ}{M}$ (par définition de R_F). Cette suite est donc également convergente dans $(\overset{\circ}{M},R_F)$; d'où le théorème .

Remarque 3 : Le lemme de trivialisation montre que si (M,F) est un modèle de type 1 de classe C^2 tel que ρ_θ est triviale, $\overset{\circ}{F}$ est une fibration de $\overset{\circ}{M}$ sur S^1. De fait il semble possible de montrer que dans tous les cas $\overset{\circ}{M}$ est fibré sur S^1, en généralisant à cette situation le théorème de Tischler (cf [22]).

IV - CROISSANCE DES FEUILLETAGES PRESQUE SANS HOLONOMIE.-

Si $C(\mathbb{N})$ est l'ensemble des applications croissantes de $\mathbb{N}$ dans lui-même, deux éléments f_1 et f_2 de $C(\mathbb{R}^+)$ qui "prolongent" un même élément f de $C(\mathbb{N})$ ont même type de croissance. Ceci nous permet de comparer les éléments de $C(\mathbb{N})$ avec ceux de $C(\mathbb{R}^+)$ et vice-versa.

Par ailleurs, si G est un groupe de type fini, S un système de générateurs de G et G_n l'ensemble des éléments de G de longueur inférieure ou égale à n (par rapport à S), on appelle fonction croissance de G (par rapport à S) la fonction $\gamma(n) = \text{card}(G_n)$ pour $n \in \mathbb{N}$. Cette définition s'étend sans difficulté au cas d'un quotient de G par un sous-groupe H, même non distingué. De plus, on sait que le type de croissance de la fonction croissance de G (et donc de la fonction croissance de G/H) est indépendant de S (cf [8]) et que γ est à croissance polynomiale si G est abélien (cf proposition 3.6 de [23]).

Nous allons maintenant démontrer le théorème 4, au § 1 pour les modèles de type 2 et au § 2 pour les modèles de type 1. Il sera facile ensuite d'établir le théorème 5 au § 3.

1 - *Croissance des modèles de type 2.*

Soient L une variété compacte de classe C^1, $q : M \to L$ une fibration (orientable) de fibre compacte K et $\mathcal{F}$ un feuilletage tangent à ∂M, transverse à la fibration. D'après le paragraphe 1.8 de [2], on sait que $(M,\mathcal{F})$ est complètement déterminé par une représentation φ du groupe fondamental de L dans le groupe des homéomorphismes (respectant l'orientation) de K. On pose $G = \text{Im}\,\varphi$ et on note G_x le sous-groupe d'isotropie de G au point $x \in K$. De plus, si on identifie K avec la fibre au-dessus du point base de L et si on désigne par F_x la feuille de $\mathcal{F}$ passant par $x \in K$, la restriction q_x de q à F_x est un revêtement de F_x sur L de fibre G/G_x. Alors si on choisit des métriques riemanniennes sur les variétés compactes M et L telles que q_x soit une isométrie locale pour tout x, on obtient le lemme ci-dessous :

Lemme 8.- Pour tout $x \in K$, *on a* $\mathrm{croiss}(F_x) = \mathrm{croiss}(G/G_x)$.

Démonstration.- Soit $\tilde{L}_\psi$ le revêtement de L associé au sous-groupe $\mathrm{Ker}\ \psi$ de $\pi_1(L)$. Pour tout x, on a un diagramme commutatif de revêtements (où p_x est un revêtement galoisien de groupe G_x):

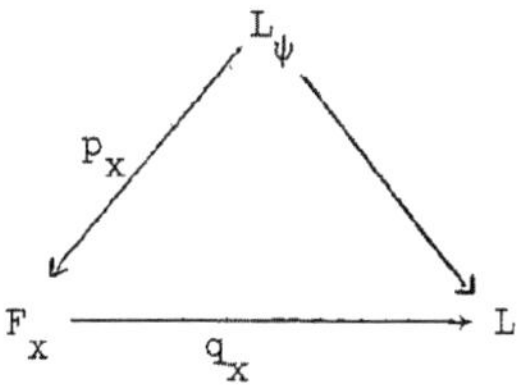

La proposition 2 de [11] montre que $\mathrm{croiss}(\tilde{L}_\psi) = \mathrm{croiss}(G)$, mais la démonstration de cette proposition se généralise sans problème au cas des revêtements non galoisiens et on obtient $\mathrm{croiss}(F_x) = \mathrm{croiss}(G/G_x)$.

Démonstration du théorème 4 (pour les modèles de type 2).-

Si (M,F) est un modèle de type 2, le groupe G est abélien d'après le lemme 5. Par suite G/G_x est abélien pour tout x, de rang inférieur ou égal à $b_1(L)$. D'après [23], le groupe G/G_x est donc à croissance polynomiale de degré inférieur ou égal à $b_1(L)$ et d'après le lemme 8, il en est de même pour le feuilletage F.

2 - *Croissance des modèles de type 1.*

Soit (M,F) un modèle de type 1 de classe C^2. On munit l'intérieur $\overset{\circ}{M}$ de M des structures S_F et R_F (cf. théorème 2). En procédant comme dans [18] (p. 110-111), on montre qu'il existe un flot $\{\overset{\circ}{\psi}_t\}_{t\in\mathbb{R}}$ de C^2-difféomorphismes de $\overset{\circ}{M}$ qui laisse globalement invariant le feuilletage $\overset{\circ}{F}$. En particulier, toutes les feuilles de $\overset{\circ}{F}$ sont difféomorphes et si j est l'injection de la feuille F_o au point base a dans $\overset{o}{M}$, on a :

i) l'homomorphisme $j_* : \pi_1(F_o,a) \to \pi_1(\overset{o}{M},a)$ est injectif.

De même, on montre que le revêtement $(\tilde{M},\pi)$ associé au sous-groupe $Im(j_*)$ de $\pi_1(\overset{o}{M},a)$ est un revêtement galoisien tel que si $\tilde{F}$, $\tilde{\vartheta}_t$ et $\tilde{R}_F$ sont les images réciproques de $\overset{o}{F}$, $\overset{o}{\vartheta}_t$ et R_F, on a :

ii) $\tilde{M}$ est difféomorphe à $F_o \times \mathbb{R}$ et, t étant une variable réelle, $\tilde{F}$ est défini par l'équation $dt = 0$;

iii) $\tilde{\vartheta}_t$ est l'application $\tilde{\vartheta}_t(x,s) = (x,s+t)$ pour $(x,s) \in F_o \times \mathbb{R}$;

iv) le groupe G est automorphismes de $(\tilde{M},\pi)$ est abélien de type fini et s'identifie à un sous-groupe dénombrable de $\{\tilde{\vartheta}_t\}_{t\in\mathbb{R}}$.

(A) Afin de simplifier les calculs à venir, on a intérêt à bien choisir la métrique initiale R sur M. Pour cela si $\partial M \neq \emptyset$, soit $\{L_j\}_{j\in J}$ l'ensemble fini des composantes connexes de ∂M. Pour tout j, il existe un voisinage tubulaire W_j de L_j difféomorphe à $L_j \times [0,1]$ tel que L_j s'identifie à $L_j \times \{1\}$ et que le champ N soit tangent au facteur $[0,1]$. Alors, si q_j est la projection de W_j sur L_j et si les W_j sont deux à deux disjoints, il existe une métrique riemannienne R sur M telle que, pour tout j, la restriction de q_j à toute feuille F_j du feuilletage induit par F dans $\overset{o}{W}_j$ soit une isométrie locale de F_j sur L_j.

<u>Lemme</u> 9.-

i) Si N *est sortant [resp. rentrant] sur* L_j, $\overset{o}{\vartheta}_t$ *est une isométrie de* $\overset{o}{W}_j$ *sur* $\overset{o}{\vartheta}_t(\overset{o}{W}_j) \subset \overset{o}{W}_j$ *pour tout* $t \geq 0$ *[resp.* $t \leq 0$*] ;*

ii) Pour $\nu > 0$, *il existe* $\eta > 0$ *tel que :*

$$\eta^{-1} ||v|| \leq ||(\vartheta_t)_*(v)|| \leq \eta||v||$$

pour tout $t \in [0,\nu]$ *et tout* $v \in T(\overset{o}{M})$.

Démonstration.-

a) Si $\mathcal{N}$ est sortant sur L_j, par exemple, on a $\varphi_t(\overset{\circ}{W}_j) \subset \overset{\circ}{W}_j$ pour tout $t \geq 0$. De plus pour $v \in T(\overset{\circ}{W}_j)$, on a $||(\varphi_t)_*(v)|| = ||v||$ par définition de R [resp. $R_{\mathcal{F}}$] si v est tangent à $\mathcal{F}$ [resp. $\mathcal{N}$]. D'où la propriété (i).

b) D'après (i), il existe une sous-variété compacte (à bord) N de $\overset{\circ}{M}$ telle que $\varphi_t(\overset{\circ}{M}-N)$ soit contenue dans $W = \bigcup_{j \in J} W_j$ pour tout $t \in [0,\nu]$. La propriété (ii) découle immédiatement de la compacité de N et $[0,\nu]$.

<u>Lemme</u> 10.- *Si* $\partial M \neq \emptyset$ *, la variété* $(\overset{\circ}{M}, R_{\mathcal{F}})$ *est à croissance linéaire.*

<u>*Démonstration.*</u>- Soit $W = \bigcup_{j \in J} W_j$. Comme $\overset{\circ}{M} - \overset{\circ}{W}$ est un compact de volume fini, le type de croissance de $\overset{\circ}{M}$ est égal au type de croissance de W. Or ce dernier est linéaire d'après le lemme 9(i).

(B) Soit θ la courbe intégrale fermée de $\mathcal{N}$ introduite au lemme 6 et soit $|\theta|$ sa longueur. Si le point base a de M appartient à θ, il existe $g_\theta \in G$ tel que $g_\theta(a,t) = (a,t+|\theta|)$ pour tout t. De plus si G_θ est le groupe engendré par g_θ et si $\hat{M} = \tilde{M}/G_\theta$, on a le diagramme commutatif de revêtements galoisiens (p étant un revêtement de groupe G/G_θ) :

(Δ) :

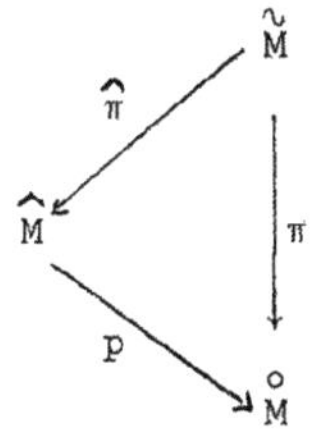

Enfin, $\hat{M}$ étant munie des structures induites $\hat{\mathcal{F}}$ et $\hat{R}_{\mathcal{F}}$, les trois revêtements du diagramme (Δ) sont injectifs sur chaque feuille du feuilletage correspondant et leur restriction à une telle feuille est un isomorphisme.

On note $\tilde{d}$ [resp. $\hat{d}$] la distance associée à $\tilde{R}_F$ [resp. $\hat{R}_F$] sur $\tilde{M}$ [resp. $\hat{M}$].

<u>Lemme</u> 11.- *Il existe un système* S *de générateurs de* G *et une constante positive* k *tels que si* $g \in G$ *et* $y \in \tilde{M}$ *vérifient la relation* $\tilde{d}(y,g(y)) \leq kn$ *, l'élément* g *est de longueur inférieure ou égale à* n *par rapport à* S.

<u>*Démonstration*</u>.- a) Considérons M munie de la métrique initiale R. Il existe un revêtement galoisien $(\bar{M},\bar{\pi})$ de M de groupe G tel que $\tilde{M}$ est l'intérieur de $\bar{M}$ et π la restriction de $\bar{\pi}$ à $\hat{M}$.

Si $\bar{M}$ est munie de la métrique image réciproque $\bar{R}$ et de la distance associée $\bar{d}$, on introduit, par analogie avec la démonstration du lemme 2 de [8], les éléments suivants :

ε le diamètre de M ;

$\bar{M}_\varepsilon$ la boule de rayon ε et centre $\tilde{a} = (a,0) \in \tilde{M} \subset \bar{M}$;

S l'ensemble fini des éléments h de G tels que

$$h(\bar{M}_\varepsilon) \cap \bar{M}_\varepsilon \neq \emptyset \text{ ;}$$

$$\ell = \inf_{g \in G-S} \bar{d}(\bar{M}_\varepsilon, g(\bar{M}_\varepsilon)) .$$

Comme dans [8], on montre alors que S est un système de générateurs de G et que si, $g \in G$ et $y \in \bar{M}$ sont tels que $\bar{d}(y,g(y)) \leq \ell n$, g est de longueur inférieure ou égale à n par rapport à S.

b) Par ailleurs, d'après la relation (Λ) introduite dans la démonstration du théorème 2, on a $\lambda\bar{R} \leq \tilde{R}_F$ sur $\tilde{M}$ et donc $\sqrt{\lambda}\,\bar{d} \leq \tilde{d}$. Le lemme découle donc immédiatement de (a) si on pose $k = \ell/\sqrt{\lambda}$.

Lemme 12.- La variété $(\tilde{M}, \tilde{R}_F)$ *est à croissance polynomiale et on a :*

$$\deg(\tilde{M}) = \begin{cases} \deg(G) + 1 & \text{si } \partial M \neq \emptyset \ ; \\ \deg(G) & \text{si } \partial M = \emptyset \ . \end{cases}$$

Démonstration.- Si $\partial M = \emptyset$, le lemme 11 n'est rien d'autre que la proposition 2 de [11] ; supposons donc ∂M non vide.

a) Pour $r \in \mathbb{R}^+$, on note $B(r)$ [resp. $\tilde{B}(r)$] la boule de rayon r, et centre $a \in \overset{\circ}{M}$ [resp. $\tilde{a} \in \tilde{M}$] ; π_r la restriction de π à $\tilde{B}(r)$ et $N(r) : \overset{\circ}{M} \to \mathbb{N}$ la fonction "multiplicité" de π_r définie par $N(r)(x) = \text{card } \{\pi_r^{-1}(x)\}$ pour $x \in \overset{\circ}{M}$. Si w est la forme volume sur $(\overset{\circ}{M}, R_F)$ on peut calculer le volume de $\tilde{B}(r)$ par la relation :

$$(V) : \text{vol } \tilde{B}(r) = \int_{B(r)} N(r)\, w \ .$$

Or si γ_S est la fonction croissance de G (par rapport à S) on voit que, d'après le lemme 11 on a $N(kn) \leq \gamma_S(2n)$ pour $x \in B(kn)$ donc aussi :

i) $\text{vol } \tilde{B}(kn) \leq \gamma_S(2n) \text{ vol } B(kn)$ pour tout $n \in \mathbb{N}$

d'après la formule (V).

b) De plus, soient $x \in B(r)$, σ une géodésique minimale reliant x au point base a et $\tilde{\sigma}$ le relèvement de σ en un point $\tilde{x} \in \pi_r^{-1}(x)$. Soit $\tilde{a}_\sigma$ l'extrémité de $\tilde{\sigma}$; si $g \in G$ est de longueur inférieure ou égale à n, et si $K = \sup_{h \in S} \tilde{d}(\tilde{a}, h(\tilde{a}))$, on a :

$$\tilde{d}(\tilde{x}, g(\tilde{x})) \leq \tilde{d}(\tilde{x}, \tilde{a}_\sigma) + \tilde{d}(\tilde{a}_\sigma, g(\tilde{a}_\sigma)) + d(g(\tilde{a}_\sigma), g(\tilde{x}))$$

$$\tilde{d}(\tilde{x},g(\tilde{x})) \leq 2r + Kn \quad \text{et} \quad \tilde{d}(\tilde{a},g(\tilde{x})) \leq 3r + Kn \ .$$

Par suite pour tout $n \in \mathbb{N}$, il vient :

$$\gamma_S(n) \leq N(4Kn)\ (x) \quad \text{pour} \quad x \in B(Kn) \ ,$$

Et finalement la formule (V) implique la relation

ii) $\gamma_S(n)$ vol $B(Kn) \leq$ vol $\hat{B}(4Kn)$.

Les formules (i) et (ii) donnent la croissance de M en fonction de la croissance de G et $\overset{\circ}{M}$; d'où le résultat annoncé, compte tenu du lemme 10.

Remarque 4.- Un calcul tout à fait analogue montre que $(\hat{M},\hat{R}_F)$ est également à croissance polynomiale avec

$$\deg(\hat{M}) \begin{cases} \deg(G/G_\theta) + 1 & \text{si } \partial M \neq \emptyset \\ \\ \deg(G/G_\theta) & \text{si } \partial M = \emptyset \ . \end{cases}$$

Ⓒ Pour en finir il nous faut comparer la croissance de F_o avec celles de $\tilde{M}$ et $\hat{M}$.

Pour ce faire, on introduit R_t, pour $t \in \mathbb{R}$, la métrique induite par $\tilde{R}_F$ sur $F_t = F_o \times \{t\} \in \tilde{\mathcal{F}}$ et d_t la distance associée. On note $D(r)$ le disque de rayon r et centre $\tilde{a} = (a,o)$ dans F_o ; $\hat{D}(r) = D(r) \times [0, |\theta|]$ et q la projection de $\tilde{M}$ sur F_o .

Lemme 13.- Si $\partial M = \emptyset$, *on a* croiss$(F_o) \geq$ croiss$(\hat{M})$.

Démonstration.-

a) Soient y_1 et y_2 deux points d'une feuille $F_t \in \tilde{\mathcal{F}}$. D'après le lemme 9 (ii), on a la relation :

$$d_t(y_1,y_2) \leq \eta\, d_o(q(y_1),q(y_2)) \quad \text{si} \quad t \in [0, |\theta|] .$$

De plus si $\tilde{\Theta} = \pi^{-1}(\theta)$, on a $g(\tilde{\Theta}) = \tilde{\Theta}$ pour tout $g \in G$ et $g_\theta(\tilde{\theta}) = \tilde{\theta}$ pour toute composante connexe $\tilde{\theta}$ de $\tilde{\Theta}$ et par suite si y_1 et y_2 appartiennent également à $\tilde{\Theta}$ on a également :

$$d_t(y_1,y_2) \leq \eta\, d_o(q(y_1),q(y_2)) \quad \text{pour tout} \quad t \in \mathbb{R} .$$

b) D'après la structure de F, il est facile de voir qu'il existe $\varepsilon > 0$ tel que $\pi(\hat{D}(\varepsilon)) = M$. On fait jouer à $\hat{D}(\varepsilon)$ le rôle de $\bar{M}_\varepsilon$ dans la démonstration du lemme 11 et on définit de la même manière S et ℓ. Alors si $k = \inf(\ell, |\theta|)$ et si $y \in \tilde{B}(kn)$, on montre, comme au lemme 2 de [8], qu'il existe $g \in G$ de longueur inférieure ou égale à n (par rapport à S) tel que $y \in g(\hat{D}(\varepsilon))$.

Par ailleurs il existe $m \in \mathbb{Z}$ tel que $|m| \leq n$ et $g_\theta^m(y) \in F_o \times [0, |\theta|]$. Donc si $K_o = \sup_{h \in S} d_o(\tilde{a}, qh(\tilde{a}))$, on obtient $d_o(\tilde{a},q\, g_\theta^m g(\tilde{a})) \leq \eta\, K_o(n + |m|)$ et finalement :

$$d_o(\tilde{a},q\, g_\theta^m(y)) \leq 2\, \eta\, K_o\, n + \eta\, \varepsilon.$$

c) Posons $2\, \eta\, K_o\, n + \eta\, \varepsilon = \alpha\, n + \beta$. D'après (b), on a

$$g_\theta^m(y) \in \hat{D}(\alpha n + \beta) \quad \text{et donc} \quad \tilde{B}(kn) \subset g_\theta^m\, \hat{D}(\alpha n + \beta).$$

De plus d'après le lemme 9 (ii), il existe $\mu > 0$ tel que $\mathrm{vol}\, \hat{D}(r) \leq \mu\, \mathrm{vol}\, D(r)$ pour $r \in \mathbb{R}^+$. D'où, en fin de compte, on a :

$$\mathrm{vol}\, \tilde{B}(kn) \leq (2n+1)\, \mu\, \mathrm{vol}\, D(\alpha n + \beta) ,$$

et connaissant les types de croissance de $\tilde{M}$ et $\hat{M}$, on en déduit que l'on a $\mathrm{croiss}(F_o) \geq \mathrm{croiss}(\hat{M})$. D'où le lemme.

Revenons au cas où ∂M n'est pas vide, et L_j étant une composante connexe de ∂M, soient W_j le voisinage tubulaire de L_j choisi au début du paragraphe (2A) et V_j une composante connexe de $\tilde{W}_j = \pi^{-1}(\mathring{W}_j)$. Pour tout j, il existe $k_j > 0$ et une isométrie locale γ_j de $\mathring{W}_j$ définie par

$$\gamma_j(y) = \mathring{\varphi}_{\pm k_j}(y) \quad \text{pour} \quad y \in \mathring{\varphi}_{\mp k_j}(\mathring{W}_j)$$

suivant que N est rentrant ou sortant sur L_j. Le relèvement de γ_j à $\tilde{M}$ définit une isométrie locale $\tilde{\gamma}_j$ de V_j et J étant fini, il existe $K > 0$ tel que pour tout $j \in J$ et pour tout $y \in F_t \cap V_j$, on a :

$$(\tilde{\gamma}) : \quad \tilde{\gamma}_j(y) \in F_t \quad \text{et} \quad d_t(y, \tilde{\gamma}_j(y)) \leq K \ .$$

Nous pouvons montrer maintenant que le résultat du lemme 13 est valable sans restriction.

<u>Lemme</u> *14.- Dans tous les cas, on a* $\text{croiss}(F_o) \geq \text{croiss}(\hat{M})$.

<u>*Démonstration*</u>.- Soit N une sous-variété compacte de $\mathring{M}$ telle que $d(\mathring{W}_j - N, \mathring{M} - W_j) \geq k_j$ pour tout $j \in J$. La variété $\tilde{N} = \pi^{-1}(N)$ est un revêtement connexe de N auquel le lemme précédent s'applique moyennant quelques adaptations évidentes.

a) Pour procéder de façon analogue au lemme 13, on choisit alors $\varepsilon > 0$ tel que $\pi(\hat{D}(\varepsilon)) \supset N$ et on définit comme précédemment. S, ℓ, K_o. En outre, on pose $\quad k = \inf\limits_{j \in J} (\ell, |\theta|, k_j)$.

Alors pour $y \in \tilde{B}(kn) \cap V_j \cap F_t$ on a $\tilde{d}(y, \tilde{M} - V_j) \leq kn$ donc il existe $m \in \mathbb{Z}$ tel que $|m| \leq n$; $z = \tilde{\gamma}_j^m(y) \in \tilde{N} \cap V_j$ et $d_t(y,z) \leq K|m| \leq Kn$. Finalement il existe $s \in \mathbb{N}$ tel que $\tilde{d}(\tilde{a},z) \leq k\,s\,n$.

b) Par ailleurs le point z appartenant à la même feuille F_t que y d'après la relation $(\tilde{\gamma})$ il existe $m' \in \mathbb{Z}$ tel que $|m'| \leq sn$ (et même $|m'| \leq n$) et $g_\theta^{m'}(z) \in F_o \times [0,|\theta|]$. Par suite z appartenant également à $\tilde{N}$ on a en procédant comme au lemme 13 (b) :

$$d_o(\tilde{a}, q\, g_\theta^{m'}(z)) \leq nK_o(sn + |m'|) \leq 2\, n\, K_o\, sn.$$

Bien plus d'après $(\tilde{\gamma})$, $g_\theta^{m'}(y) \in F_o \times [0, |\theta|]$ et

$$d_o(q\, g_\theta^{m'}(z), q\, g_\theta^{m'}(y)) \leq nK\, |m| \leq n\, K\, n\, .$$

En fin de compte, on obtient :

$$d_o(\tilde{a}, q\, g_\theta^{m'}(y)) \leq n\, (2\, K_o\, s + K)\, n\, ;$$

et il n'y a plus alors qu'à conclure comme au lemme 13.

Démonstration du théorème 4 (pour les modèles de type 1).-

a) Avec les notations précédentes, on a $\hat{\pi}(\hat{D}(r) \subset \hat{B}(r + |\theta|)$ pour tout r et donc, la restriction de $\hat{\pi}$ à $\hat{D}(r)$ étant injective presque partout, on obtient d'après la formule (V) :

$$\mathrm{vol}\, \hat{D}(r) \leq \mathrm{vol}\, \hat{B}(r + |\theta|)\, .$$

Par ailleurs, d'après le lemme 9 (ii), il existe une constante μ' telle que $\mu'\, \mathrm{vol}\, D(r) \leq \mathrm{vol}\, \hat{B}(r + |\theta|)$ et à l'aide du lemme 14 on en déduit que

$$\mathrm{croiss}(F_o) = \mathrm{croiss}(\hat{M})\quad .$$

b) Toute feuille F de $\mathcal{F}$ est isométrique à une feuille F_t de $F_o \times [0,|\theta|]$ et le lemme 9 (ii) montre que alors $\mathrm{croiss}(F_t) = \mathrm{croiss}(F_o) = \mathrm{croiss}(\hat{M})$. Bref le feuilletage $\mathcal{F}$ est à croissance polynomiale avec $\deg(\mathcal{F}) = \deg(\hat{M})$.

c) Pour finir, si $\partial M \neq \emptyset$, $\deg(F) = 1$ équivaut à $G = G_\theta$ donc $\hat{M} = \mathring{M}$ et $\hat{F} = \mathring{F}$ c'est-à-dire $\mathring{F}$ est une fibration de $\mathring{M}$ sur S^1.

D'où le théorème.

3 - *Cas général : démonstration du théorème 5.*

En combinant les théorèmes 1 et 4, on voit immédiatement que tout feuilletage (M,F) presque sans holonomie, de classe C^2, est à croissance polynomiale.

Enfin si F possède une feuille compacte, toute feuille de F appartient à un modèle de type 1 et donc $\deg(F) = 1$ si et seulement si toutes les feuilles de F sont propres (cf. théorème 4). D'où le théorème.

V - FEUILLETAGES A CROISSANCE D'ORDRE INFERIEUR OU EGAL A UN.-

Pour démontrer les théorèmes 6 et 7, nous aurons à utiliser une autre définition de la croissance des feuilles d'un feuilletage : la croissance transversale qui peut être définie elle en classe C^0.

Pour cela, soient Ω un recouvrement bidistingué régulier d'axe X de (M,F,N) et Γ_X le pseudo-groupe d'holonomie de F réduit à X.

Pour $F \in F$ et $a \in F \cap X$, on appelle fonction croissance transversale de F en a la fonction croissance g_a^T de la trajectoire T_a de Γ_X en a pour la métrique δ (cf introduction de III). Bien sûr, le type de croissance de g_a^T est indépendant du point a et du système (fini) de générateurs de Γ_X et même on montre que g_a^T a même type de croissance que la fonction croissance géométrique g_a de F en a (cf. [17] § 4 et paragraphe I.a) .

Rappelons encore deux définitions de [7]; les notations étant celles introduites au début du paragraphe III, on dit qu'une feuille F capte une feuille F' s'il existe $f \in \Gamma_X$, $a \in D_f \cap F$ et $a' \in D_f \cap F'$ tels que $f(a) = a$ et $\lim_{n\to+\infty} f^n(a') = a$. Evidemment, si F capte une feuille F' elle capte toutes les feuilles voisines du même côté que F'. Si une feuille se capte elle-même, on l'appelle une feuille-ressort ; une feuille-ressort est exceptionnelle ou localement dense, une feuille qui capte et qui n'est pas une feuille-ressort est semi-propre.

Démonstration du théorème 6.- Si F est une feuille-ressort, il existe, avec les notations précédentes, un élément $g \in \Gamma_X$ tel que $a \in D_g$ et $g(a) = a'$. Soit alors Γ_Y le pseudo-groupe engendré par les restrictions de f et g à un voisinage compact Y de a contenant a'. La relation d'équivalence associée possède un minimal compact unique contenant le point a donc non fini. En procédant exactement comme dans [16], on montre alors que la croissance de la trajectoire de a par Γ_Y est exponentielle. Il en est de même a fortiori pour la trajectoire de a par Γ_X et par suite F est à croissance exponentielle. D'où le théorème.

L'utilisation des feuilles-ressort permet d'affiner la classification des germes de feuilletages au voisinage d'une feuille compacte de [6]. (voir aussi théorème 8 de [17]).

Lemme 15.- *Soit* F *une feuille compacte d'un feuilletage* $\mathcal{F}$ *de classe* C^2 *à croissance non-exponentielle. Les deux conditions ci-dessous sont équivalentes :*

i) F *est isolée à droite [resp. à gauche] dans l'ensemble des feuilles compactes de* $\mathcal{F}$ *;*

ii) F *capte les feuilles voisines à droite [resp. à gauche].*

Démonstration.- En effet, si F est isolée à droite et ne capte pas les feuilles à droite, le germe de $\mathcal{F}$ à droite de F est non dérivable au sens de [6]. En particulier, les feuilles du feuilletage induit par $\mathcal{F}$ dans tout voisinage tubulaire à droite de F (suffisamment petit) sont denses. Donc il existe des feuilles-ressort et d'après le théorème 6, $\mathcal{F}$ est à croissance exponentielle. Ce qui démontre le lemme.

Enfin soit F une feuille semi-propre qui capte une feuille F' différente de F à l'aide de l'homéomorphisme local f. Si Σ est un système de générateurs de Γ_X , on pose $\Sigma' = \Sigma \cup \{f\}$ et on note γ_Σ et $\gamma_{\Sigma'}$ les fonctions croissance transversale (par rapport aux systèmes de générateurs Σ et Σ') des feuilles F et F' aux points a et a' respectivement. On a le résultat suivant :

<u>*Lemme 16*</u>*.- Si* F *est de classe* C^2*, on a l'inégalité :*

$$\gamma_{\Sigma'}(n) \geq \sum_{p=0}^{n} \gamma_{\Sigma}(p) \quad \textit{pour tout } n \in \mathbb{N} .$$

En particulier, si F *est polynomiale de degré* m *, on a* $\text{croiss}(F') \geq \text{croiss}(x^{m+1})$.

<u>*Démonstration*</u>.- Si F est de classe C^2 et si F est semi-propre, on sait (cf. proposition 2) qu'il existe un voisinage V de a dans X tel que tout raccourci g en a soit défini sur V; et on choisit le point a' dans V.

De plus, soient T et T' les trajectoires de Γ_X en a et a' munies des métriques δ et δ' (cf paragraphe III). Si $x \in T$ est tel que $\delta(a,x) = r \leq m$, il existe un raccourci g en a tel que $g(a) = x$ et pour tout $s \in \{0,1,\dots,m-r\}$ on a $\delta'(a',g \circ f^s)(a')) \leq m$.

Enfin si h est un raccourci en a tel que $h(a) \neq g(a)$, on a $(g \circ f^s)(a') \neq (h \circ f^t)(a')$ pour tout $(s,t) \in \mathbb{N}^2$ d'après la semi-propreté de F. La relation annoncée s'ensuit sans peine.

<u>*Démonstration du théorème 7*</u>.- Soit (M,F) un feuilletage de classe C^2 dont la croissance est d'ordre inférieur ou égal à 1. D'après le théorème de Plante, F ne possède pas de minimal exceptionnel. De plus,

a) si toutes les feuilles de F sont denses, F est un modèle de type 1 à croissance linéaire (cf. théorème 4) ;

b) si F possède une feuille compacte, on peut supposer, quitte à couper le long de certaines feuilles compactes, que M est une variété à bord et que les feuilles compactes de F sont contenues dans ∂M.

Toute feuille non compacte de F est alors captée par une feuille du bord d'après le lemme 15 et par suite elle est à croissance linéaire d'après le lemme 16. Bien plus d'après le théorème 6 et le lemme 16, l'holonomie de toute feuille non compacte est triviale. Bref (M,F) est un modèle de type 1 (avec bord), et le reste du théorème 7 découle du théorème 4.

VI - EXEMPLES DE CROISSANCES.-

Tout feuilletage F de classe C^∞ de $V_2 \times [-e,+e]$ (où $e = + 10$) qui est C^∞-tangent au bord définit par identification des deux composantes du bord un feuilletage de classe C^∞ de $V_2 \times S^1$ que nous désignerons encore par le même symbole F. Aussi, pour démontrer les théorèmes 8 et 9 nous allons construire les feuilletages correspondants sur $V_2 \times [-e,+e]$.

Or pour $a \in V_2$, il existe un homomorphisme de $\pi_1(V_2,a)$ sur le groùpe libre à deux générateurs. Donc si G est un sous-groupe à deux générateurs de $\mathrm{Diff}^\infty_+([-e,+e])$, il existe une représentation de $\pi_1(V_2,a)$ dans $\mathrm{Diff}^\infty_+([-e,+e])$ dont l'image est G. Par la méthode décrite au paragraphe 1.8 de [2], on obtient alors un feuilletage de $M = V_2 \times [-e,+e]$, transverse au facteur $[-e,+e]$ et C^∞-tangent au bord si les générateurs de G sont C^∞-tangent à l'identité en $(-e)$ et $(+e)$.

En outre, d'après le lemme 8, on sait que pour tout $x \in [-e,+e]$, on a $\mathrm{croiss}(F_x) = \mathrm{croiss}(G/G_x)$ si G_x est le sous-groupe d'isotropie de G en x et si F_x est la feuille de F passant par le point (a,x). De plus, il est facile de voir que pour tout $n \in \mathbb{N}$, la fonction croissance de G/G_x en n est égale au nombre de raccourcis en x de longueur inférieure ou égale à n dont les images sont deux à deux distinctes.

En fait, les feuilletages que nous allons construire pour démontrer le théorème 8 correspondent aux diverses étapes de construction du feuilletage (1.b) sur $V_2 \times [-e,+e]$ de [5], et pour le théorème 9, ce sont les feuilletages (1.b) et (1.c) de [5] eux-mêmes. Nous nous contenterons donc ici de donner quelques idées sur les méthodes utilisées.

Démonstration du théorème 8.- Soit f un C^∞-difféomorphisme de $\mathbb{R}$ de support $[-e,+e]$ tel que :

(f_1) $f(-1) = +1$ et $f(x) > x$ pour $x \in (-e,+e)$;

(f_2) f est C^∞-tangent à l'identité en $(-e)$ et $(+e)$.

On note encore f la restriction de f à $[-e, +e]$.

a) Le feuilletage $\mathcal{F}_1$.

Il correspond au groupe G_1 engendré par f. Le groupe G_1 étant isomorphe à $\mathbb{Z}$, le feuilletage $\mathcal{F}_1$ est évidemment linéaire, presque sans holonomie.

b) Le feuilletage $\mathcal{F}_2$.

Soit h l'homothétie de rapport e et soit G_2 le sous-groupe de $\mathrm{Diff}_+^\infty([-e,+e])$ engendré par les restrictions à $[-e,+e]$ de f et $f_2 = h^{-1} \circ f \circ h$. On peut faire les remarques suivantes :

i) le groupe G_2 est à croissance exponentielle. En effet, pour toute suite finie $(i_1,\dots,i_m)$ à valeurs dans $\{0,1\}$, les éléments de la forme

$$f \circ f_2^{i_1} \circ f \circ f_2^{i_2} \circ \dots \circ f \circ f_2^{i_m}$$

sont deux à deux distincts, car si k et ℓ sont deux tels éléments correspondants respectivement aux suites finies $(i_1,\dots,i_m)$ et $(j_1,\dots,j_n)$, on a :

$$k(0) = (f^m \circ f_2^{i_m})(0) \quad \text{et} \quad (0) = (f^n \circ f_2^{j_n})(0)$$

Donc si $k = \ell$ on a $m = n$; $i_m = j_n$ et par récurrence $i_s = j_s$ pour tout $s \in \{1,\dots,m\}$. Par suite la fonction croissance de G_2 domine la fonction 2^m.

ii) le feuilletage $\mathcal{F}_2$ défini par G_2 possède deux feuilles compactes $F(0)$ et $F'(0)$ seulement. De plus, tout raccourci au point $+1$ étant de la forme f^n, $n \in \mathbb{Z}$, la feuille $F(+1) = F_1$ est à croissance linéaire.

Par ailleurs, son groupe d'holonomie engendré par le germe de f_2 est cyclique non trivial.

iii) enfin, toutes les autres feuilles de $\mathcal{F}_2$ sont difféomorphes, les sous-groupes d'isotropie correspondants de G_2 sont tous isomorphes et donc toutes ces feuilles ont même type de croissance. Or tout raccourci en 0 est de la forme $(f^p \circ f_2^q)$ avec $(p,q) \in \mathbb{Z}^2$ et $(f^p \circ f_2^q)(0) = (f^r \circ f_2^s)(0)$ implique $(p,q) = (r,s)$. Par suite F_0 est à croissance quadratique et il en est de même pour $\mathcal{F}_2$ ainsi que prévu.

c) Cas général : le feuilletage $\mathcal{F}_n$.

Soient $f_3 = h^{-1} \circ f_2 \circ h$; $\hat{f}_3 = f \circ f_3 \circ f^{-1}$ et $g_3 = \hat{f}_3 \circ f_2$. On définit $\mathcal{F}_3$ à l'aide du groupe G_3 engendré par les restrictions à $[-e,+e]$ de f et g_3.

Les autres feuilletages sont alors définis par récurrence. Le difféomorphisme f_{n-1}, $\hat{f}_{n-1}$ et g_{n-1} étant définis à l'étape $(n-1)$, on pose pour $n > 3$:

$$f_n = h^{-1} \circ f_{n-1} \circ h \ ;$$

$$\hat{f}_n = f^{-n+2} \circ f_n \circ f^{n-2} \quad \text{et} \quad g_n = \hat{f}_n \circ g_{n-1} \ .$$

Il est facile de vérifier que le feuilletage $\mathcal{F}_n$ défini par le groupe G_n engendré par f et g_n possède les propriétés voulues.

Démonstration du théorème 9.-

i) Remarquons que la suite $\{g_n\}_{\substack{n\in\mathbb{N}\\ n\geq 3}}$ définie dans la démonstration du théorème 7 converge vers un homéomorphisme g de $\mathbb{R}$ à support dans $[-e,+e]$. Cet homéomorphisme peut être rendu C^∞ par une définition appropriée des difféomorphismes f_n .

Le groupe G_b engendré par les restrictions de f et g définit le feuilletage (1,b) de [5]. Par suite, on sait que pour tout $p \in \mathbb{N}$, la feuille $F_{e^{1-p}}$ est propre, d'holonomie non triviale et que toutes les autres feuilles sont difféomorphes, de même croissance et partout denses.

Or il est facile de voir que la feuille $F_{e^{1-p}}$ est isomorphe à la feuille $F(p)$ de $\mathcal{F}_n$ pour $n > p$. Elle est donc à croissance polynomiale de degré p.

Il nous reste alors à vérifier que F_0 est à croissance exponentielle et pour cela, il suffit de remarquer que si $(i_1,\dots,i_m)$ est une suite finie à valeurs dans $\{-1, +1\}$, les éléments de G_b de la forme

$$f \circ g^{i_1} \circ \dots \circ f \circ g^{i_m} \circ f^{-m}$$

sont des raccourcis en 0 de longueur $3m$ dont les images en 0 sont deux à deux distinctes. Alors la fonction croissance de G_b/G_0 domine la fonction 2^m. Bref le feuilletage $\mathcal{F}_b$ défini par G_b a les propriétés voulues.

ii) La construction de $\mathcal{F}_c$, est plus complexe, nous renvoyons donc au feuilletage (1,c) de [5].

REFERENCES

[1] P.R. DIPPOLITO - *The structure of codimension one folioations II : Reeb Stability (preprint).*

[2] A. HAEFLIGER - *Variétés feuilletées* ; Ann. Scuola Norm. Sup. Pisa, 16 (1964), 367-397.

[3] G. HECTOR - *Sur un théorème de structure des feuilletages de codimension 1.* Thèse, Strasbourg, 1972.

[4] G. HECTOR - *Sur les feuilletages presque sans holonomie.* C.R. Acad. Sc. Paris, 274 (1972), 1703-1706.

[5] G. HECTOR - *Quelques exemples de feuilletages. Espèces rares* - (à paraître aux Ann. Inst. Fourier, 26 (1) (1975).

[6] G. HECTOR - *Classification cohomologique des germes de feuilletages (preprint).*

[7] C. LAMOUREUX - *Sur quelques phénomènes de captage.* Ann. Inst. Fourier, 23 (4) (1973), 229-243.

[8] J. MILNOR - *A note on curvature and the fondamental group.* J. of Diff. Geometry, 2 (1968), 1-7.

[9] J. MILNOR - *Growth of finitely generated solvable groups.* J. of Diff. Geometry, 2 (1968), 447-449.

[10] R. MOUSSU - *Sur les feuilletages de codimension 1.* Thèse, Orsay, 1971.

[11] R. MOUSSU et F. PELLETIER - *Sur le théorème de Poincaré-Bendixson.* Ann. Inst. Fourier, 24 (1)(1974), 131-148.

[12] J. PALIS et C. PUGH - *Fifty problems in dynamical systems.* Dynamical Systems, Warwick 1974, Lecture Notes n° 468, 345-353.

[13] J. PLANTE - *Asymptotic properties of foliations.* Comm. Math. Helv., 47 (1972), 449-456.

[14] J. PLANTE - *A generalization of the Poincaré-Bendixson theorem for foliations of codimension one.* Topology, 12 (1973), 177-181.

[15] J. PLANTE - *On the existence of exceptional minimal sets in foliations of codimension one.* J. of Diff. Eq., 15 (1974), 178-194.

[16] J. PLANTE - *Measure preserving pseudogroups and a theorem of Sacksteder.* Ann. Inst. Fourier, 25(1), (1975), 237-249.

[17] J. PLANTE - *Foliations with measure preserving holonomy.* Ann. of Math. 102 (1975), 327-361.

[18] G. REEB - *Sur certaines propriétés topologiques des variétés feuilletées.* Act. Sc. et Ind. Hermann, Paris, 1952.

[19] B.L. REINHART - *Foliated manifolds with bundle-like metrics.* Ann. of Math. 69 (1959), 119-131.

[20] R. SACKSTEDER - *Foliations and pseudo-groups.* Amer. J. of Math. 87 (1965), 79-102.

[21] R. SACKSTEDER et A. SCHWARTZ - *Limit sets of foliations.* Ann. Inst. Fourier, 15 (2) (1965), 201-214.

[22] D. TISCHLER - *On fibering certain foliated manifolds over* S^1 . Topology, 9 (1970) 153-154.

[23] J. WOLF - *Growth of finitely generated solvable groups and curvature of Riemannian manifolds.* J. of Diff. Geometry, 2 (1968), 421-446.

Gilbert Hector
IMPA
Rua Luiz de Camões, 68
Rio de Janeiro

Adresse habituelle :
Université des Sciences et Techniques de Lille
U.E.R. de Mathématiques
BP 36
59650 - VILLENEUVE D'ASCQ

SUR LA THEORIE DES FEUILLETAGES ASSOCIEE AU REPERE MOBILE : CAS DES FEUILLETAGES DE LIE

Edmond FEDIDA

1 . INTRODUCTION

On connaît depuis longtemps la dualité entre l'algèbre des formes différentielles extérieures et l'algèbre de Lie des champs de vecteurs.

Cette dualité permet en particulier de présenter la théorie (différentielle) des structures feuilletées, soit dans le langage des systèmes de PFAFF complètement intégrables, soit dans celui des champs de vecteurs en involution.

En outre, on a réservé, au sein de la théorie des structures feuilletées, une place importante aux groupes de transformations de Lie qui correspondent à des systèmes de champs de vecteurs en involution, associés à une algèbre de Lie. Il n'est donc pas étonnant que cette théorie possède un pendant, qui n'est autre que la théorie des feuilletages associée au repère mobile. Il est utile de préciser rapidement cette "analogie" pour situer convenablement notre problème.

.../...

I

SYSTEME DYNAMIQUE
ou
GROUPE DE TRANSFORMATIONS DE LIE

On se donne une variété V^n de dimension n munie d'un système de champs de vecteurs X_i, et des constantes de structures $C_{i\,jk}$ d'un groupe de Lie tels que : $[X_i, X_j] = C_{ijk} X_k$

II

REPERE MOBILE

On se donne une variété V^n de dimension n munie d'un système de formes de Pfaff ω_i et des constantes de structures C_{ijk} d'un groupe de Lie avec les relations :

$$d\omega_i = C_{ijk}\, \omega_j \wedge \omega_k$$

Propriété commune : à chacune des deux structures est associé l'algèbre de Lie $\mathcal{G}$ de constantes de structures C_{ijk} .

Soit G un groupe de Lie connexe d'algèbre de Lie $\mathcal{G}$ et soit $\tilde{G}$ (resp. $\tilde{V}^n$) le revêtement universel de G (resp. de V^n) .

Propriété 1

à la structure correspond une application de $V^n \times \tilde{G}$ dans V^n :

$$V^n \times \tilde{G} \to V^n$$
$$(x,t) \to x_t$$

telle que :

1) $x_e = x$
2) $(x_t)_{t'} = x_{tt'}$

Propriété 1'

à la structure correspond une application de $\tilde{V}^n \times G$ dans G

$$\tilde{V}^n \times G \to G$$
$$(x,t) \to t_x$$

telle que :

1) $t_{x_o} = t$; $x_o \in V^n$ fixé
2) $t'(t_x) = (tt')_x$

(loi du repère mobile)

Propriété 2

si rang $(X_i) = n$ en tout point, V^n est un espace homogène de groupe G .

Propriété 2'

si rang $(\omega_i) = n$ en tout point, on a une injection locale de V^n dans G .

Propriété 3

la loi de groupe définit dans V^n des trajectoires formant une partition de V^n . La trajectoire est unique dans le cas de la propriété 2 .

Propriété 3'

la loi de cogroupe définit une partition de $\tilde{V}^n$ formée des "classes constantes" de l'application de $\tilde{V}^n$ dans G $\tilde{V}^n$ étant le revêtement universel de V^n . Les classes sont ponctuelles dans le cas de la propriété 2'.

Propriété 4

les trajectoires de $\tilde{G}$ portent une structure d'espace homogène associé à G .

Propriété 4'

les classes de $\tilde{V}^n$ possèdent une structure transverse de groupe associé à G .

CAS PARTICULIERS INTERESSANTS

1) Les X_i forment un système de rang constant en tout point : Les trajectoires constituent alors un feuilletage de V^n ; les feuilles sont des espaces homogènes.

1') Les ω_i forment un système de rang constant en tout point . Les classes forment un feuilletage de V^n dont la structure transverse est modelée sur celle d'un sous espace de G .

2) $X_1 \ldots X_\ell$ forment un système de rang constant et une sous algèbre de l'algèbre engendrée par les X_i .
Les trajectoires ont une structure d'espace homogène.

2') On se donne un sous anneau J de rang constant complètement intégrable de l'anneau des ω_i .
Le feuilletage associé, a une structure transverse modelé sur un espace homogène de G . De tels feuilletages méritent le nom de feuilletages homogènes : la structure géométrique des feuilletages homogènes associés en groupe affine de R (feuilletages linéaires) ainsi que des théorèmes d'existence des feuilles compactes sont donnés dans [4] et [5]

3) Les X_i sont linéairement indépendant en chaque point .
Les trajectoires ont alors $\tilde{G}$ pour revêtement universel.

3') Les ω_i sont linéairement indépendants en tout point :
Le feuilletage associé a une structure transverse modelée sur G . Un tel feuilletage mérite alors le nom de feuilletage de Lie

Le cas 3') qui retient notre attention dans cet article, a une importance particulière, du fait qu'on peut toujours s'y ramener, à partir de n'importe quelle situation mentionnée dans la deuxième colonne ; l'équivalent d'une telle propriété fait curieusement défaut dans la théorie des groupes de transformations de Lie.

Plus précisément, en langage vectoriel, une structure II est donnée sur V^n par une 1-forme ω à valeurs dans une algèbre de Lie $\mathcal{G}$.

de dimension q , qui vérifie l'équation de Maurer-Cartan

$$(1) \quad d\omega + \frac{1}{2}[\omega,\omega] = o$$

on peut alors "désingulariser" le feuilletage défini par ω sur V^n en considérant sur le fibré principal trivial $V^n \times G$, la forme de connexion Ω induite par ω ; l'équation (1) implique alors que Ω est une forme de connexion plate . En particulier Ω est une 1-forme sur $V^n \times G$ à valeurs dans $\mathfrak{G}$ surjective vérifiant l'équation de Maurer-Cartan : $d\Omega + \frac{1}{2}[\Omega,\Omega] = o$. On est ainsi ramené sur $V^n \times G$ au cas 3').

Considérons donc, une algèbre de Lie réelle $\mathfrak{G}$ de dimension q et G un groupe de Lie connexe ayant $\mathfrak{G}$ pour algèbre de Lie.

Soit V^n une variété différentiable et $\omega : TV^n \to \mathfrak{G}$ une forme différentielle de degré 1 sur V^n à valeurs dans $\mathfrak{G}$ ayant les propriétés suivantes :

i) $d\omega + \frac{1}{2}[\omega,\omega] = o$ (condition de Maurer-Cartan)

ii) $\omega : T_x(V^n) \to \mathfrak{G}$ est surjective pour tout $x \in V^n$.

Dans ces conditions ω détermine un feuilletage $\mathcal{F}$ de codimension q de V^n ; et on dira que $\mathcal{F}$ est un $\mathfrak{G}$-feuilletage de Lie de V^n .

Les feuilletages de Lie sont mentionnés dans (7) et abordés sur des cas particuliers dans [6] , [9] et [11] .

Exemples.

1) Une forme de connexion plate sur un fibré principal E de groupe G détermine un $\mathfrak{G}$-feuilletage de Lie de E .

2) Une forme de Pfaff fermée et sans singularités sur V^n , détermine un R-feuilletage de Lie de V^n , où R est l'algèbre de Lie triviale de dimension 1 .

3) MOLINO a établi récemment [8] les résultats suivants :

i) Soient V^n une variété compacte connexe de dimension n , $\mathcal{F}$ un feuilletage de codimension q , sur V^n , transversalement parallélisable (i.e il existe q champs de vecteurs feuilletés $X_1 \dots X_q$, qui engendrent en tout point un supplémentaire de l'espace tangent à la feuille). Dans ce cas, les adhérences des feuilles de $\mathcal{F}$ sont les fibres d'une fibration $\Pi : V^n \to W$. De plus il existe une algèbre de Lie $\mathfrak{g}$ telle que $\mathcal{F}$ induit sur chaque fibre de Π un $\mathfrak{g}$-feuilletage de Lie. On peut noter qu'un $\mathfrak{g}$-feuilletage de Lie est transversalement parallélisable.

ii) D'une manière générale, soient V^n une variété connexe de dimension n, $\mathcal{F}$ un feuilletage de codimension q sur V^n, de type transitif (i.e. l'ensemble des champs de vecteurs feuilletés complets est transitif sur V^n). Dans ce cas les adhérences des feuilles de $\mathcal{F}$ sont les fibres d'une fibration $\Pi : V^n \to W$; $\mathcal{F}$ induisant sur chaque fibre un $\mathfrak{g}$-feuilletage de Lie. On remarquera que tout feuilletage défini par une fibration est de type transitif.

iii) Soient V^n une variété compacte connexe de dimension n, un feuilletage de codimension q sur V^n admettant une métrique quasi-fibrée [12]. Sur le fibré E_T des repères transverses orthonormés, on a un feuilletage relevé $\mathcal{F}_T$, dont les adhérences des feuilles définissent une fibration $\Pi_T : E_T \to W_T$ et $\mathcal{F}_T$ induira sur les fibres un $\mathfrak{g}_T$-feuilletage de Lie.

2 . QUELQUES PROPRIETES DES FEUILLETAGES DE LIE

Soit $\mathcal{F}$ un **$\mathfrak{g}$-feuilletage de** Lie de codimension q sur V^n ; défini par la donnée d'une 1-forme ω sur V^n à valeurs $\mathfrak{g}$ de rang q en chaque point et vérifiant $d\omega + \frac{1}{2}[\omega,\omega] = o$. Soient G un groupe de Lie connexe de dimension q, d'algèbre de Lie $\mathfrak{g}$, α la forme de Maurer-Cartan de G, et Ω la forme de connexion sur le fibré principal trivial $V^n \times G$ égale à $\omega - \alpha$ sur $V^n \times \{e\}$.

Cette forme détermine un $\mathcal{G}$-feuilletage de Lie Φ de $V^n \times G$ ayant les propriétés suivantes :

i) Φ est invariant par les translations à droite de G

ii) Φ est transverse aux fibres $\{x\} \times G$, $x \in V^n$

iii) Φ est transverse aux fibres $V^n \times \{g\}$, $g \in G$ est induit sur chacune de ces fibres le feuilletage $\mathcal{F}$

iv) la projection $V^n \times G \to G$ induit une submersion sur chaque feuille de Φ

v) la projection $V^n \times G \to V^n$ induit un revêtement galoisien sur chaque feuille de Φ . Si $h : \Pi_1(V^n) \to G$ est l'homomorphisme d'holonomie de la connexion Ω , le groupe d'automorphismes de ce revêtement est isomorphe à $h(\Pi_1(V^n))$.

On en déduit alors facilement :

Proposition 1 : Soit $\mathcal{F}$ un $\mathcal{G}$-feuilletage de Lie d'une variété V^n :

i) le groupe G opère transitivement sur l'espace des feuilles.

ii) $\mathcal{F}$ est sans holonomie.

Soit $W^n \subset V^n \times G$ une feuille de Φ et soient $p : W^n \to V^n$ et $f : W^n \to G$ les projections de W^n sur V^n et G .

Proposition 2 : Le feuilletage $\mathcal{F}^* = p^*\mathcal{F}$ de W^n est déterminé par la submersion $f : W^n \to G$.

Théorème 1 : Si V^n est compacte la projection $f : W^n \to G$ est une fibration localement triviale.

Démonstration

Si V^n est muni d'un $\mathcal{G}$-feuilletage de Lie $\mathcal{F}$, il existe un recouvrement ouvert de V^n , $\{U_i\}_{i \in I}$ et pour chaque $i \in I$, une submersion

$f_i : U_i \to G$. Si $U_i \cap U_j \neq \emptyset$, il existe une application localement constante de $U_i \cap U_j$ dans G telle que :

(1) $$f_i = g_{ij} f_j \ .$$

On munit G d'une métrique η invariante à gauche, et on choisit un sous-fibré τ de $T(V^n)$ transverse à $\mathfrak{F}$. Par les f_i on relève η à τ en une métrique $\mathfrak{T}$: la condition de compatibilité est donnée par (1). On complète alors $\mathfrak{T}$ en une structure riemanienne $\mathfrak{M}$ de V^n par une métrique "horizontale" assujettie à la seule condition qu'en tout point $x \in V^n$, l'espace tangent à la feuille est orthogonal pour $\mathfrak{M}$, à l'espace transverse. Par suite $\mathfrak{F}$ possède une métrique quasi-fibrée au sens de [12]. V^n étant compacte, $\mathfrak{M}$ est complète. La métrique $p^* \mathfrak{M}$ induite sur W^n est alors quasi-fibrée et complète et se projette par f sur une métrique invariante à gauche sur G . On sait dans ce cas [7] qu'on peut prolonger indéfiniment les trajectoires orthogonales aux fibrés de f . Ce résultat acquis, on procède par récurrence sur la dimension de la base ; la question étant locale, on peut supposer que la base est $\mathbb{R}^q$.
Pour $q = 1$, Reeb a montré [11] que W^n est difféomorphe à $W_{n-1} \times \mathbb{R}$. Supposons le résultat vrai pour $q-1$. On considère l'image réciproque par f de $\mathbb{R}^{q-1}$ dans W^n , soit T , qui est de codimension 1 . On fibre $\mathbb{R}^q$ par les parallèles à $\mathbb{R}^{q-1}$ et W^n par les images réciproques T_t $(t \in \mathbb{R})$. En considérant les trajectoires orthogonales des T_t , on se trouve ramené au cas de Reeb, et W^n est donc difféomorphe à $T \times \mathbb{R}$; l'hypothèse de récurrence appliquée à la restriction de f à T permet de conclure.

Corollaire 1 : Si V^n est compacte, l'espace quotient de $\mathcal{F}$ est $\frac{G}{h(\Pi_1(V^n))}$ c'est à dire une Q-variété au sens de [1].

En effet la projection $f : W^n \to G$ est surjective et $h(\Pi_1(V^n))$ opère sur W^n et G d'une manière équivariante.

Corollaire 2 : Si V^n est compacte le feuilletage $\mathcal{F}$ possède la propriété du prolongement des homotopies (au sens de [6]).

La démonstration est analogue à celle donnée par GODBILLON [6] dans le cas où $\mathcal{G} = \mathbb{R}$.

Soit $\bar{K}$ l'adhérence de $h(\Pi_1(V^n)) = K$ dans G ; l'image réciproque par f , d'une classe à gauche suivant $\bar{K}$ dans W^n , reprojetée sur V^n compacte, est une sous variété fermée de V^n, adhérence d'une feuille de $\mathcal{F}$ et réunion de feuilles ; si bien qu'on a une décomposition de V^n en un feuilletage (non de Lie sauf si $\bar{K}$ est distingué dans G) dont l'espace quotient est $G/\bar{K}$; et, dans chacune des feuilles fermées de ce feuilletage, on a un feuilletage de Lie de groupe $\bar{K}$ dont toutes les feuilles sont partout denses ; d'où :

Théorème 2 : Les adhérences des feuilles de $\mathcal{F}$ sont les fibres d'une fibration $\Pi : V^n \to G/\bar{K}$. De plus il existe une algèbre de Lie $\mathcal{K}$ telle que $\mathcal{F}$ induit sur chaque fibre de Π un $\mathcal{K}$-feuilletage de Lie à feuilles partout denses.

On a déjà remarqué dans le paragraphe 1 que Molino a étendu ce résultat à d'autres classes de feuilletages.

3 . PROBLEMES D'EXISTENCE

Soit V^n une variété différentielle munie d'un $\mathcal{G}$-feuilletage de Lie $\mathcal{F}$ de codimension q , défini par le couple (ω , $\mathcal{G}$)

$\mathcal{F}$ admet une métrique quasi-fibrée ; le produit intérieur associé étant noté $\langle\ \rangle$.

Une base de $\mathcal{G}$ étant fixée, soient $\omega_1, \ldots, \omega_q$ les composantes de ω par rapport à cette base ; les q champs de vecteurs sur V^n ,

$X_1, \ldots, X_q$ définis par :

$$\langle \omega_i , X_j \rangle = \delta_{ij}$$

trivialise le fibré normal associé à $\mathcal{F}$; d'où :

Proposition : L'existence d'un feuilletage de Lie de codimension q sur une variété ouverte V^n est équivalente à celle d'une submersion de V^n dans $\mathbb{R}^q$

Démonstration

Soit $Sub(V^n, \mathbb{R}^q)$ (resp. Sec Tq V^n) l'ensemble des submersions de V^n dans $\mathbb{R}^q$ (resp. l'ensemble des sections du fibré des q-plans de V^n) on sait [10] que :

$$Sub(V^n ; \mathbb{R}^q) \neq \phi \Longleftrightarrow \text{Sec } T_q V^n \neq \phi$$

Si V^n admet donc un feuilletage de Lie de codimension q, Sec $T_q V^n \neq \phi$ et par suite $Sub(V^n , \mathbb{R}^q) \neq \phi$. Réciproquement il est clair que $Sub(V^n ; \mathbb{R}^q) \neq \phi$ entraine $Sub(V^n , G) \neq \phi$ pour tout groupe de Lie connexe G de dimension q. On a vu au paragraphe 2, qu'à tout $\mathfrak{g}$-feuilletage de Lie sur une variété compacte V^n correspond une fibration d'un revêtement galoisien de V^n sur un groupe de Lie G, d'algèbre de Lie $\mathfrak{g}$.
Il est donc naturel d'étudier des obstructions à l'existence de fibrations d'une variété donnée dans un groupe de Lie G ; le théorème qui suit précise un résultat intéressant obtenu dans cette direction, dans le cas où G est compact. On en déduit en particulier qu'une sphère ne peut pas être fibrée (sauf cas triviaux) sur un groupe de Lie. Je remercie vivement le Professeur J.P. SERRE, pour ses précieuses remarques concernant la démonstration du théorème suivant :

Théorème : Si un espace fibré E a pour base B, un groupe de Lie compact connexe et simplement connexe de dimension > 0, et si la cohomologie des fibres F de E est nulle en dimension assez grande, alors le 3ème nombre de Betti de E est non nul.

La démonstration de ce théorème est donnée dans [3]

Remarque

Les nombres de Betti de E ne sont pas les mêmes que ceux de B x F .

Corollaire : Si une variété V^n compacte et simplement connexe admet un $\mathfrak{g}$- feuilletage de Lie, alors on a :

$$H^3 (V^n , \mathbb{R}) \neq 0 .$$

En particulier il n'existe pas de feuilletage de Lie sur les sphères.

En effet l'existence d'un $\mathfrak{g}$-feuilletage de Lie sur V^n compacte et simplement connexe entraîne que V^n est un fibré de base G , un groupe de Lie compact , connexe et simplement connexe d'algèbre de Lie $\mathfrak{g}$;
d'où $H^3(V^n,\mathbb{R}) \neq 0$.
Si V^n est une sphère, on a donc nécessairement $V^n = S_3$

4 . CONSTRUCTION DES $\mathfrak{g}$-FEUILLETAGES DE LIE

Un procédé classique de construction de $\mathfrak{g}$-feuilletages de Lie est le suivant :

La donnée d'homomorphisme de $\Pi_1(V^n)$ dans G , un groupe de Lie d'algèbre de Lie $\mathfrak{g}$, détermine un fibré principal $E \to V^n$ de groupe G et une connexion plate sur E , donc un $\mathfrak{g}$-feuilletage de Lie de E.
Dans l'une des deux hypothèses ci-dessous.

a) $\Pi_1(V^n)$ est un groupe libre

b) $\Pi_1(V^n)$ est un groupe abélien libre, G compact et $H^*(G)$ sans torsion.

ONIŠČIK a montré [9] qu'on obtenait ainsi une connexion plate sur $V^n \times G$, donc un $\mathfrak{g}$-feuilletage de Lie de $V^n \times G$, qui induit en général sur V^n un $\mathfrak{g}$-feuilletage de Lie avec "singularités".

Exemple

Soit G un groupe de Lie connexe semi-simple compact et simplement connexe. On sait que tout groupe de Lie semi-simple connexe, contient un sous groupe libre H engendré pas deux éléments tels que $\bar{H} = G$.

Soient $h : H \to G$ l'injection de H dans G, et M une variété compacte telle que $\Pi_1(M) = H$.

h induit alors sur $M \times G$ un feuilletage de Lie de groupe G, dont toutes les feuilles sont partout denses et simplement connexes.

Terminons sur les problèmes suivants :

i) faire une étude analogue pour les feuilletages homogènes (voir paragraphe 1)

ii) trouver des obstructions à l'existence d'un $\mathfrak{g}$-feuilletage de Lie sur une variété donnée (par exemple si $\mathfrak{g} = \mathbb{R}$ et V^n compacte, on doit avoir $H^1(V^n, \mathbb{R}) \neq 0$.)

iii) caractériser les variétés admettant un $\mathfrak{g}$-feuilletage de Lie (pour $\mathfrak{g} = \mathbb{R}^p$, les variétés compactes sur les fibrés sur T^p [13]).

R E F E R E N C E S

1 BARRE R. : De quelques aspects de la théorie des Q-variétés différentielles et analytiques ; Thèse, Strasbours (Juin 1972)

2 FEDIDA E. : Feuilletages du plan , Feuilletage de Lie ; Thèse, Strasbourg (Octobre 1973)

3 FEDIDA E. : Sur l'existence des feuilletages de Lie ; C.R. Acad. Sc. Paris § 278 p. 835- 837 (18 Mars 1974)

4 FURNESS P.M.D. et FEDIDA E. : Transversally affine foliations ; Glasgow Math. J. (à paraître)

5 FURNESS P.M.D. et FEDIDA E. : Sur le feuilletage linéaires : (à paraître aux C.R.A.S.).

6 GODBILLON C. : Feuilletages ayant la propriété du prolongement des homotopies Ann. Inst. Fourier 17,2 (1967) p. 219 - 260

7 HERMANN R. : On the différential geometry of foliations ; Ann. Math. 72 (1960) p. 445 - 457

8 MOLINO P. : Feuilletages transversalement parallélisables et feuilletages de Lie ; (à paraître aux C.R.A.S.)

9 ONISCK A.L. : Some Concepts and applications of non abelian cohomology theory ; Trudy Mosk. Math. obsc. 17 (1967) p. 45-88.

10 PHILLIPS A. : Submersions of open manifolds ; Topology 6 (1967) p. 171-206

11 REEB G. : Sur certaines propriétés topologiques des variétés feuilletées Act. Sc. et Ind. Hermann Paris (1952)

12 REINHART B. : Foliated manifolds with bundlelike metrics. Ann. of Math. , 69 (1959) p. 119 - 131

13 TISHLER D. : On fibering certain foliated manifold over S^1 Topology 9 (1970)

ADRESSE : Edmond FEDIDA, Département de Mathématiques, Faculté des Sciences Université de Dakar . DAKAR -(Sénégal)

On the Index Of Isolated Closed Tori

R. J. Knill

There seems to be a rather universal acceptance of an integer or rational valued index for isolated closed orbits of flows based on the Poincaré method of sections. There is less understanding of how to assign a reasonable integer valued index to an isolated closed torus of an $\mathbb{R}^n$ action, $n > 1$. Although Bobylev and Krasnosel'skii [1] have given such an index, it is not well defined in the generality in which they give it (see paragraph 4).

The purpose of this paper is to explain how to obtain an integer valued index for an isolated closed torus of an $\mathbb{R}^n$ action, $1 < n$, and to relate it to previously defined indexes.

§1. The integer valued index of an isolated closed torus.

Let M be a smooth (that is C^∞) m-manifold and let g be a smooth action of $\mathbb{R}^n$ on M, some $n \geq 1$. Then g is a smooth map of $M \times \mathbb{R}^n$ into M such that for x in M and vectors u and v in $\mathbb{R}^n$, $g(g(x,u),v) = g(x,u+v)$ and $g(x,0) = x$.

Let F be an isolated closed torus of g. Its set of periods form a lattice in $\mathbb{R}^n$. Let p be one such period not equal to zero. Then by "isolated" is meant that there is a neighborhood N of F in M and a positive ε such that for $x \in N$ and $v \in \mathbb{R}^n$ such that $|p - v| < \varepsilon$, $g(x,v) = x$ if and only if $x \in F$ and $v = p$. By "torus" we mean to assume in addition that for $x \in F$, $F = g(x \times \mathbb{R}^n)$, and F is compact. Equivalently, $\mathbb{R}^n$ modulo the subgroup of periods of a base point x of F is a torus which is mapped diffeomorphically onto F by the map induced by g.

Let z_0 be a point of F, fixed once and for all. Let $\alpha: N_0 \to U \times B \subset \mathbb{R}^n \times \mathbb{R}^{m-n}$ be a chart about z_0 such that for $u \in U$, $\alpha^{-1}(u,0) = g(z_0, p + u)$. This means α maps $N_0 \cap F$ diffeomorphically onto $U \times \{0\}$. It follows that $\alpha^{-1}(\{0\} \times B)$ is transverse to F, and we can assume it is transverse to every orbit, by taking B

smaller if necessary. To simplify notation denote a point $\alpha^{-1}(x,y)$ by (x,y). Then z_0 would be written $(0,0)$.

1. Lemma. *We may choose* U_1 *and* B_1 *so small that there are unique smooth maps*

$$\lambda : U_1 \times B_1 \to U' \qquad w : U_1 \times B_1 \to B$$

such that for every $(x,y) \in U_1 \times B_1$,

$$g(x,y,\lambda(x,y)) = (x, w_x(y))$$

where $w_x(y) = y$ *if and only if* $y = 0$, *and where* $\lambda(x,0) = p$ *for* $x \in U_1$.

Proof: Write $g(x,y,\lambda) = (g_1(x,y,\lambda), g_2(x,y,\lambda))$ whenever $g(x,y,\lambda) \in U \times B$. We may suppose that U_1, B_1 and $\varepsilon > 0$ are so small that $g(x,y,\lambda) \in U \times B$ whenever $(x,y) \in U_1 \times B_1$ and $|p - \lambda| < \varepsilon$. Let $T_\varepsilon = \{\lambda \in \mathbb{R}^n : |p - \lambda| < \varepsilon\}$ and let

$$G : U_1 \times B_1 \times B_1 \times T_\varepsilon \to \mathbb{R}^m$$

be defined by

$$G(x,y,w,\lambda) = (g_1(x,y,\lambda) - x, g_2(x,y,\lambda) - w).$$

Then the Jacobian

$$\left.\frac{\partial(G_1,G_2)}{\partial(w,\lambda)}\right|_{(0,p)} = \det \begin{pmatrix} 0 & \dfrac{\partial g_1}{\partial \lambda} \\ -\mathrm{id} & \dfrac{\partial g_2}{\partial \lambda} \end{pmatrix} \Bigg|_{(0,p)}$$

and this determinant is nonzero since $\left.\dfrac{\partial g_1}{\partial \lambda}\right|_{(0,p)}$ is nonsingular.

Furthermore $G(x,0,0,p) = (0,0)$ for $x \in U_1$. By the implicit function theorem the neighborhoods U_1 and B_1 may be chosen so small that there exist unique smooth functions $\lambda(x,y)$ and $w_x(y)$ satisfying

$$G(x,y,w_x(y),\lambda(x,y)) = (0,0), \ \lambda(x,0) = p.$$

Equivalently,

$$g(x,y,\lambda(x,y)) = (x, w_x(y)).$$

We may further assume that $\lambda(x,y) \in T_\varepsilon$. But as remarked earlier, ε may be chosen so small that $g(x,y,\lambda) = (x,y)$ for $(x,y,\lambda) \in U \times B \times T_\varepsilon$ if and only if $y = 0$ and $v = p$. It follows that $w_x(y) = y$ if and only if $y = 0$. We have already the condition $\lambda(x,0) = p$, for $x \in U_1$.

In explanation of this lemma there is for every x in U_1 a unique map $w_x : B_1 \to B$ such that for some $\lambda \in T$, $g(x,y,\lambda) = (x,w_x(y))$. This map w_x is the "cross sectional" map defined by the Poincaré method of sections. As in the case of n = 1, it depends on the period chosen, as well as other choices such as of the local representation $U \times B$ of a neighborhood of z_0 in M, and even the choice of z_0. We want to show that the fixed point index $i(w_x,B_1)$ is independent of all choices (including the choice of $z_0 \in F$), but excluding the choice of period p. For this purpose we relate it to the θ^* homomorphism of [4]. For a definition of fixed point index see [5] or [7].

§2. The θ_* homomorphism

The θ_* homomorphism is a homomorphism which is defined under the following circumstances. There is given an ANR X and a normal Hausdorff space T which in the case considered in paragraph 1 are the manifold M and the group $\mathbb{R}^n$, respectively. Let V be an open subset of $X \times T$. There is given a continuous function $g: V \to X$ and a closed subset S of $X \times T$ all of whose points $(x,t) \in S$, satisfy $g(x,t) = x$. S is <u>isolated</u> by V in the sense that

$$S = \{(x,t) \in V \mid g(x,t) = x\}.$$

We do not assume that either X or T is compact but we do assume that g(V) has compact closure. Let K be a compact subset of X containing the projection of S into X. In the case under consideration in paragraph 1, K would equal F and S would equal $F \times \{p\}$.

We let T_0 be any subspace of T such that $X \times T_0$ is disjoint from S. (In case $X = M$, $T = \mathbb{R}^n$, and $S = F \times \{p\}$ we would take $T_0 = R^n \backslash \{p\}$).

From [4] there is a homomorphism

$$\theta_*(g) : H_*(T,T_0) \to H_*(K) \qquad \text{(rational coeff.)}$$

which satisfies the following properties

(i) (Normalization) If $V = X \times T$, if T is compact and T_0 is empty then there is a Lefschetz type formula (in terms of cap product) for the composition, $\Lambda_*(g)$, of the homomorphisms

$$H_*(T) \xrightarrow{\theta_*(g)} H_*(K) \xrightarrow{i_*} H_*(X)$$

where i_* is induced by the inclusion.

(ii) (Additivity) If S is a union of disjoint closed subsets S_1 and S_2 and if V_1 and V_2 are disjoint open neighborhoods in V of S_1 and S_2, respectively, then

$$\theta_*(g) = \theta_*(g|V_1) \oplus \theta_*(g|V_2) .$$

(iii) (Naturality in T) Suppose that T' is another normal Hausdorff space and $\tau : T' \to T$ is a continuous function. Let $V' = (\mathrm{id} \times \tau)^{-1}(V)$ and let $g' : V' \to X$ be defined by $g(x,t') = g(x,\tau(t'))$ for $(x,t') \in X \times T'$. Let T_0' be a subspace of T' such that $\tau(T_0') \subset T_0$. Then as homomorphisms of $H_*(T',T_0')$ into $H_*(K)$, we have that

$$\theta_*(g') = \theta_*(g) \circ \tau_* .$$

(iv) (Homotopy invariance) Suppose that $h_s : V \to X$, $0 \le s \le 1$, is a homotopy such that $\bigcup_{0\le s\le 1} h_s(V)$ has compact closure and the set

$$S(h) = \{(x,t,s) \in V \times [0,1] \mid h_s(x,t) = x\}$$

is closed in $X \times T$. Let K be a compact set containing the projection of $S(h)$ into X, and let T_0 be such that $X \times T_0$ is disjoint from $S(h)$. Then

$$\theta_*(h_0) = \theta_*(h_1).$$

(v) (Commutativity) Suppose that X' is any other ANR and that there are continuous maps

$$f : V \to X' , \; r : X' \to X$$

such that $g = r \circ f$. Let

$$V' = \{(x',t) \in X' \times T' \mid (r(x'),t) \in V\}$$

let $g'(x',t) = f(r(x'),t)$ for $(x',t) \in V'$, and let

$$S' = \{(x',t) \in V' \mid g'(x',t) = x'\} .$$

The projection of S' into X' is mapped by r into K so there is a compact subset K' of X' containing this projection and mapped into K by r. In such a case

$$r_* \circ \theta_*(g') = \theta_*(g).$$

(This is a simplified form of this property as it appears in [4].)

In addition we wish to announce the following form of multiplicativity. This is a generalization common to theorem 27 of [5], and the multiplicativity property of [4]. Its proof will appear in a paper in preparation.

(vi) (Multiplicativity) Let $X = X_1 \times X_2$, $T = T_1 \times T_2$, $T_0 = T_1 \times T_{20} \cup T_{10} \times T_2$ and suppose that V has the form

$$V = V_1 \times V_2$$

where $V_1 = D_1 \times E_1$ is contained in $X_1 \times T_1$ and $V_2 \subset X_2 \times T_2$. Suppose that D_1 is connected. Let

$$g : V_1 \times V_2 \to X_1 \times X_2$$

have the form

$$g(x_1,t_1,x_2,t_2) = (g_1(x_1,t_1),g_2(x_1,x_2,t_2)).$$

Suppose that $K = K_1 \times K_2$. Then

(a) $\theta_*(g_2(x_1,\cdot,\cdot))$ is independent of $x_1 \in D_1$ (since D_1 is connected).

(b) $\theta_*(g) = \theta_*(g_1) \otimes \theta_*(g_2(x_1,\cdot,\cdot))$ as homomorphisms of $H_*(T_1,T_{10}) \otimes H_*(T_2,T_{20})$ into $H_*(K_1) \otimes H_*(K_2)$.

An application of the normalization property was given by theorem 3 of [4]. The next theorem is proven the same way.

2. Theorem. *Let* $X = T = T^n$ *be the* n *fold product of* S^1 *with itself. Let* $m : T^n \times T^n \to T^n$ *be the standard group operation,* $m(x,y) = x + y$. *Then*

$$\theta_*(m) : H_n(T^n) \to H_n(T^n)$$

is the identity map.

Let $\mathbb{Z}^n \subset \mathbb{R}^n$ be the n-fold product of the integers with themselves. Then $T^n = \mathbb{R}^n/\mathbb{Z}^n$. Let $\tau : \mathbb{R}^n \to T^n$ be the quotient map and define the canonical action of $\mathbb{R}^n$ on T^n as the function $m' : T^n \times \mathbb{R}^n \to T^n$ defined by $m'(x,v) = m(x,\tau(v))$. Let p be a lattice point (i.e. point of $\mathbb{Z}^n$). Then p is a period of the $\mathbb{R}^n$ action.

3. Lemma. $\theta_*(m') : H_n(\mathbb{R}^n, \mathbb{R}^n \setminus \{p\}) \to H_n(T^n)$ *is the composition of the isomorphisms*

$$\begin{array}{ccc} H_n(\mathbb{R}^n, \mathbb{R}^n \setminus \{p\}) & \xleftarrow[\text{excision}]{\cong} & H_n(E^n, E^n \setminus \{p\}) \\ & & \cong \downarrow m'(x,\cdot)_* \\ & & H_n(T^n, T^n \setminus \{x\}) \xleftarrow[\text{inclusion}]{\cong} H_n(T^n). \end{array}$$

Here x is any element of T^n and E^n is a small enough disk centered at p for $m(x,\cdot)$ to map E^n homeomorphically onto a neighborhood of x in T^n.

Proof: By the naturality property there is a commutative diagram

$$
\begin{array}{ccc}
H_n(\mathbb{R}^n, \mathbb{R}^n \setminus \{p\}) & \xrightarrow{\theta_*(m')} & H_n(T^n) \\
\text{excision} \uparrow \cong & & \| \\
H_n(E^n, E^n \setminus \{p\}) & \xrightarrow{\theta_*(m')} & H_n(T^n) \\
\tau^* \downarrow \cong & & \| \\
H_n(T^n, T^n \setminus \{0\}) & \xrightarrow{\theta_*(m)} & H_n(T^n) \\
\text{inclusion} \uparrow \cong & & \| \\
H_n(T^n) & \overset{id=\theta_*(m)}{=\!=\!=\!=} & H_n(T^n) .
\end{array}
$$

Furthermore the bottom two vertical arrows on the left factor also as the isomorphisms below.

$$
\begin{array}{ccccc}
 & & H_n(E^n, E^n \setminus \{p\}) & & \\
 & \overset{m'(x,\cdot)}{\swarrow} & & \overset{\tau_* = m'(0,\cdot)_*}{\searrow} & \\
H_n(T^n, T^n \setminus \{x\}) & & \xrightarrow{m(-x,\cdot)} & & H_n(T^n, T^n \setminus \{e\}) \\
 & \underset{\substack{\text{inclusion} = m(x,\cdot)_* \\ \text{(since } T^n \text{ is connected)}}}{\nwarrow} & & \underset{\text{inclusion}}{\nearrow} & \\
 & & H_n(T^n) & &
\end{array}
$$

By putting these diagrams together, one sees that $\theta_*(m')$ is as claimed.

§3. The relation of $i_*(w_x, B_1)$ to $\theta_*(g)$.

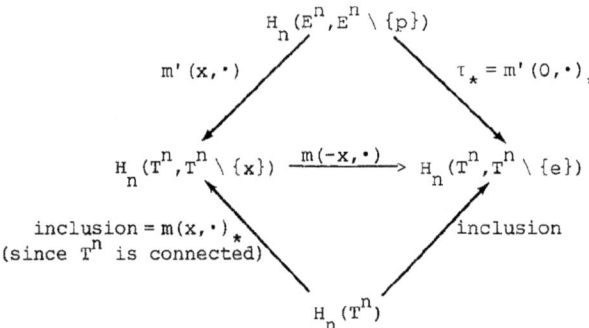

Now we are ready to relate the θ_* homomorphism to the index $i(w_x, B_1)$ defined in the first paragraph. Let the notation be as in paragraph 1. For an orientation class μ of $H_n(\mathbb{R}^n, \mathbb{R}^n \setminus \{p\})$, there is a corresponding orientation class μ_F of $H_n(F)$ (recall that F is homeomorphic to T^n), defined as the image of μ under the composition of isomorphisms

$$\begin{array}{ccc} H_n(\mathbb{R}^n,\mathbb{R}^n \setminus \{p\}) & \xleftarrow[\text{excision}]{\cong} & H_n(E^n,E^n \setminus \{p\}) & & \\ & & \Big\downarrow g(z_0,\cdot)_* & & \\ & & H_n(F, F \setminus \{z_0\}) & \xleftarrow[\text{inclusion}]{\cong} & H_n(F) \end{array}$$

where z_0 is any element in F.

4. Theorem. *For a small enough neighborhood* V *of* $F \times \{p\}$,

$$\theta_*(g|V)(\mu) = i(w_x,B_1) \cdot \mu_F .$$

Proof: In outline we will show that for the purposes of this proof we may assume that F has a tubular neighborhood N which is homeomorphic to $F \times B$. Then the analysis of paragraph 1 yields a function $w_x : B_1 \to B$ for each x in F. Let

$$g'(x,y,v) = (g_1(x,0,v), w_x(y)).$$

Then it will be shown that $g|V$ and $g'|V$ are homotopic through actions $h_s : V \to F \times B$, $0 \le s \le 1$, such that F is the only closed torus of h_s for each s. By the homotopy property

$$\theta_*(g|V) = \theta_*(g'|V) .$$

(The homotopy h_s, $0 \le s \le 1$, is defined by

$$h_s(x,y,t) = (g_1(x,sy,t), g_2(x,y,st + (1-s)\lambda(x,y))$$

where g_1 and g_2 are as in lemma 1. Then $h_1 = g|V$ and $h_0 = g'|V$. One readily checks that $h_s(x,t) = x$ if and only if $x \in F$ and $t = p$, for $(x,t) \in F \times B_1$. We can assume $V = F \times B_1$ by covering $F \times \{p\}$ with sets of the form of $U_1 \times B_1$ from lemma 1.) Represent $\mathbb{R}^n$ as $\mathbb{R}^n \times \{*\}$ where $*$ is an ideal point. Then we may write

$$g'(x,y,v) = (g_1'(x,v), g_2'(x,y,*))$$

where

$$g_1'(x,v) = g_1(x,0,v) \quad \text{for} \quad x \in F,\ v \in T_\varepsilon$$

and where

$$g_2'(x,y,*) = w_x(y).$$

By the multiplicativity property

$$\theta_*(g|V) = \theta_*(g_1') \otimes \theta_*(g_2'): H_n(\mathbb{R}^n, \mathbb{R}^n \setminus \{p\}) \otimes H_0(*)$$

$$\to H_n(F) \otimes H_0(\{0\}).$$

But after identifying $H_0(*)$ = rationals = $H_0(\{0\})$, one sees that $\theta_*(g_2')(1) = N$, and by the uniqueness of fixed point indexes [2], $N = i(w_x, B_1)$. Thus

$$\theta_*(g|V)(\mu) = i(w_x, B_1) \cdot \theta_*(g_1')(\mu).$$

Claim $\theta_*(g_1')(\mu) = \mu_F$: By the commutativity property $\theta_*(g_1')$ is a topological invariant, and topologically g_1' acts as m' so lemma 3 yields that $\theta_*(g_1')(\mu) = \mu_F$. As a consequence

$$\theta_*(g|V)\mu = i\ (w_x, B_1)\mu_F\ .$$

It remains then to show that we may assume that F has a trivial tubular neighborhood. For this we must leave the realm of $\mathbb{R}^n$ actions, that is we can no longer assume that $g(z,0) = z$, for all z, although it will still be the case that $h(z,u+v) = g(g(z,u),v)$. The reader will note that the first property was not needed in the discussion up until now. The first step is to embed M in $\mathbb{R}^m$ as a closed subset. Since the tangent bundle of F is trivial, then the normal bundle of F in $\mathbb{R}^m$ is stably trivial, so by increasing m sufficiently we may assume that the normal bundle of F in $\mathbb{R}^m$ is in fact trivial. Let $M^\#$ be an open neighborhood of M in $\mathbb{R}^m$ such that there is a smooth retraction $r : M^\# \to M$. Replace M with $M^\#$ and g with the map $g^\#$ defined by $g^\#(z,v) = g(r(z),v)$ for $z \in M^\#$ and $v \in \mathbb{R}^n$. $F \times \{p\}$ is isolated, that is, there is a neighborhood $V^\#$ of $F \times \{p\}$ in $M^\#$ such that

$$F \times \{p\} = \{(z,v) \in V^\# \mid g^\#(z,v) = z\}$$

The previous analysis assuming F had a trivial tubular neighborhood now really does apply to $g^\#|V^\#$, and so for appropriate $B_1{}^\#$,

(1) $$\theta_*(g^{\#}|V^{\#})\mu = i\ (w_x{}^{\#}, B_1{}^{\#})\mu_F\ .$$

Note that $g^{\#}|F \times \mathbb{R}^n = g|F \times \mathbb{R}^n$ so μ_F may be defined equally well in terms of $g^{\#}$ or g. Secondly, one readily sees that in the local coordinates of lemma 1,

$$g^{\#}(x,y,\lambda(x,y)) = (x,0,w_x \circ r(y))$$

on $U_1 \times (B_1{}^{\#} \cap r^{-1}(B_1))$. Therefore

$$w_x{}^{\#} = w_x \circ r\ .$$

But since r is a retraction,

$$i\ (w_x, B_1) = i\ (w_x \circ r, B_1{}^{\#})\ .$$

This reduces (1) to

(2) $$\theta_*(g^{\#}|V^{\#})(\mu) = i\ (w_x, B_1)\mu_F.$$

Finally we apply the commutativity property to g and $g^{\#}$ to get a commutative triangle

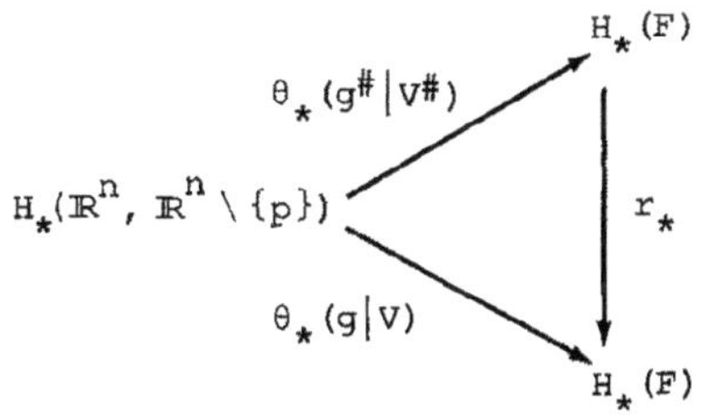

Since $r|F$ = identity this proves

(3) $$\theta_*(g|V) = \theta_*(g^{\#}|V^{\#})$$

and so the theorem is proven by (2) and (3).

5. <u>Corollary</u>. $i(w_x, B_1)$ *is independent of the choice of representation of a neighborhood of* z_0 *as a product as well as of* x.

Note that it is not claimed that $i(w_x, B_1)$ is independent of period p. For example, on $\mathbb{C}$ = complex numbers the derivative of

$y^2 - y$ at the origin is -1. Then it is possible to extend $y^2 - y$ to a diffeomorphism $f : C \to C$ which equals $y^2 - y$ near the origin and equals $-y$ near infinity. Let g be the "suspension" of f, that is, let X be the space formed from $[0,1] \times C$ by identifying points $(1,y)$ with $(0,f(y))$. For $(x,y) \in [0,1] \times C$, let $\langle x,y \rangle$ be the equivalence class in X represented by (x,y). Define $g : X \times \mathbb{R}^1 \to X$ by $g(\langle x,y \rangle,\lambda) = \langle f^n(x),t \rangle$ where $x + \lambda = n + t$, n is an integer and $0 \le t < 1$. Then $F = \{\langle x,0 \rangle : 0 \le x \le 1\}$ is an isolated closed orbit of F, with integer periods. If $p = n \ge 1$, then the map $w_0(y)$ corresponding to p is $w_0(y) = f^n(y)$. One computes that $i(f,B_1) = 1$, $i(f^2,B_1) = 3$, for a suitably small ball neighborhood B_1 of 0 in C. Thus $i(w_0,B_1)$ does depend on the period p.

§4. The index of Bobylev and Krasnosel'skii.

Bobylev and Krasnosel'skii [1], have defined an index as follows. Let X be a domain in a Banach space E, and let $g : X \times \mathbb{R}^n \to E$ be a completely continuous operator. This means that the image of g has compact closure. Suppose that there is a compact smooth connected manifold F in X such that for some neighborhood V of $F \times \{p\}$,

$$F \times \{p\} = \{(x,v) \in V \mid g(x,v) = x\}.$$

No semigroup property is assumed for g by [1]. For a point z_0 in F it is supposed that there is a smooth map Λ of a neighborhood N of z_0 into $\mathbb{R}^n$ such that Λ maps $N \cap F$ diffeomorphically onto its image and $\Lambda^{-1}(p) \cap F$ consists only of z_0. Then one says that Λ is <u>transverse</u> to F at the point z_0. Let $a_\Lambda(z) = g(z,\Lambda(z))$ for $z \in N$. Then z_0 is an isolated fixed point of a_Λ, and the topological index of this fixed point is denoted by Bobylev and Krasnosel'skii as

$$\operatorname{ind}(z_0,\Lambda) = i(a_\Lambda,N).$$

Bobylev and Krasnosel'skii claimed without proof that

6. *Suppose the smooth continuum* F *is connected. Then the number* $|\operatorname{ind}(z,\Lambda_0)|$ [*which equals* $|i(a_\Lambda,N)|$] *is the same for all* $z_0 \in F$ *and all transformations* Λ *which are transverse to* F.

Later we will give a counterexample. For $\mathbb{R}^n$ actions one can redefine the Bobylev and Krasnosel'skii index by restricting attention to a canonical choice of Λ as follows.

7. Theorem. *If* g *is a group action of* $\mathbb{R}^n$ *on* M, *then there is a canonical smooth map* Λ *of a neighborhood* N *of* z_0 *in* X *such that in the notation of theorem* 4,

$$\theta_*(g|V)(\mu) = i(a_\Lambda,N)\mu_F .$$

In particular $i(a_\Lambda,N)$ *is independent of* $z_0 \in F$.

Proof. If E is finite dimensional, then theorem 4 applies and

$$(1) \qquad \theta_*(g|V)(\mu) = i(w_x,B_1) \cdot \mu_F .$$

If E is not finite dimensional, then F has a trivial tubular neighborhood in X, and so the proof of theorem 4 applies to yield (1). It remains then to define Λ. Applying lemma 1, there is a neighborhood $N = U_1 \times B_1$ of z_0 in X such that U_1 is a neighborhood of z_0 relative to F and B_1 is identified with the open ball in some Banach space E' and such that there are unique smooth maps

$$\lambda : U_1 \times B_1 \to U' \subset \mathbb{R}^n$$

$$w : U_1 \times B_1 \to B \subset E'$$

such that for every $(x,y) \in U_1 \times B_1$,

$$g(x,y,\lambda(x,y)) = (x,w_x(y)).$$

Write $z_0 = (x_0,0)$. Let U' be a neighborhood of the origin in $\mathbb{R}^n$ such that the map $u \longmapsto g(z_0,u)$ maps U' diffeomorphically onto an image neighborhood of z_0 in F, taking 0 to z_0. We may assume that U' is contained in the image of U'. Furthermore, we may assume that every point of $U_1 \times B_1$ is of the form $g(x_0,y,v)$ for unique $y \in B$ and $v \in U'$. So we may define

$$\Lambda : U_1 \times B_1 \to \mathbb{R}^n$$

by

$$\Lambda\, g(x_0,y,v) = \lambda(x_0,y) - v .$$

Then one readily computes using the group property of g that $a_\Lambda(x,y)$ is given by

$$\begin{aligned} a_\Lambda(x,y) &= g(g(x_0,y,v),\lambda(x_0,y) - v) \\ &= g(x_0,y,\lambda(x_0,y)) \\ &= (x_0,w_{x_0}(y)). \end{aligned}$$

Thus

$$\begin{aligned} i(a_\Lambda,U_1 \times B_1) &= i(a_\Lambda|x_0 \times B_1,x_0 \times B_1) \\ &= i(w_{x_0},B_1). \end{aligned}$$

This completes the proof of the theorem.

As a *counterexample* to item 6, suppose that $X = F = T^n$ and we look at the action m' of $\mathbb{R}^n$ on T^n defined in paragraph 2. Let $p = 0$, let z_0 be a base point of T^n and let

$$f : \mathbb{R}^n \to T^n$$

be defined by $f(v) = m'(z_0,v)$. Then f is a local diffeomorphism. Let U' be a small enough neighborhood of $0 \in \mathbb{R}^n$ for $f|U'$ to be one to one, define $U = f(U')$. For any diffeomorphism $h : \mathbb{R}^n \to \mathbb{R}^n$ leaving exactly 0 fixed, let $\Lambda : U \to \mathbb{R}^n$ be defined by

$$\Lambda(m'(z_0,v)) = h(v) - v \qquad v \in U'.$$

Then

$$\begin{aligned} a(z) &= m'(z,\Lambda(z)) \\ &= m'(m'(z_0,v),h(v) - v) \\ &= m'(z_0,h(v)) \\ &= f \circ h \circ (f|U')^{-1}(z). \end{aligned}$$

By the topological invariance of fixed point indexes,

$$i(a_\Lambda,U) = i(h,U').$$

But h may be chosen with any index desired so $|i(a_\Lambda,U)|$ is not independent of the choice of Λ.

8. Theorem. *If* Λ_0 *is canonical and* $i(a_{\Lambda_0},U) = 0$, *then for any* $\Lambda : U \to \mathbb{R}^n$ *such that* Λ *is transverse to* $\mathbf{F}$,

$\operatorname{ind}(z_0,\Lambda) = 0$. *In particular if* $\operatorname{ind}(z_0,\Lambda) \neq 0$ *for any* Λ *then the closed torus is stable in the sense that if* $g^\#$ *is close enough to* g, *there is a closed torus of* $g^{\#}$ *close to* F *with period close to the period* p *of* F .

Proof: Let V be a neighborhood of the origin of a Banach space E' and let $h : V \to E$ be a transversal to F at z_0. We may choose V small enough and may choose a small enough neighborhood U' of the origin in $\mathbb{R}^n$ such that the map

$$\Gamma : U' \times V \to E$$

defined by

$$\Gamma(x,y) = g(h(y),x)$$

is a local diffeomorphism. Evidently

$$g(\Gamma(x,y),v) = \Gamma(x+v,y)$$

for x and v in a small enough neighborhood of the origin in $\mathbb{R}^n$. Suppose that U' is in fact so small.

Let U be a neighborhood of z_0 contained in $\Gamma(U' \times V)$ and let $\Lambda : U \to \mathbb{R}^n$ be a smooth map such that $\Lambda | U \cap F$ is a diffeomorphism, such that $\Lambda(z_0) = p$, where $z_0 = \Gamma(0,0)$. Let λ and w be as in theorem 4. In the notation of this proof,

$$g(\Gamma(x,y),\lambda(x,y)) = \Gamma(x,w_x(y))$$

and $\lambda(x,0) = p$ for all x in U'. Then

$$\begin{aligned} a_\Lambda(\Gamma(x,y)) &= g(\Gamma(x,y), \Lambda(\Gamma(x,y))) \\ &= g(\Gamma(0,y),x + \Lambda(\Gamma(x,y)) + \lambda(0,y) - \lambda(0,y)) \\ &= g(g(\Gamma(0,y),\lambda(0,y)),x + \Lambda\Gamma(x,y) - \lambda(0,y)) \\ &= g(\Gamma(0,w_0(y)),x + \Lambda\Gamma(x,y) - \lambda(0,y)) \\ &= \Gamma(x + \Lambda\Gamma(x,y) - \lambda(0,y),w_0(y)). \end{aligned}$$

By the multiplicativity property

$$i(a_\Lambda,U) = i(id + \Lambda\Gamma - \lambda, U') \times i(w_0,V).$$

But $i(w_0,V) = i(a_{\Lambda_0},U)$ by the proof of theorem 7. Since $i(a_{\Lambda_0},U) = 0$, then $i(a_\Lambda,U) = 0$. The stability is a consequence of the homotopy property of θ_*. This proves the theorem.

§5. Concluding comments.

By proving theorem 4, §3, we have shown that the integer $i(w_x, B_1)$ of an isolated closed orbit is independent of all choices save that of the isolated orbit and its period. In the case of $\mathbb{R}^1$ actions, Fuller [3] has gone somewhat farther in establishing that the integer valued index may be extended to dynamical systems whose closed orbits are not necessarily isolated. We feel that the analogous extension for $\mathbb{R}^n$ actions also holds. The usefulness of such an extension would be analogous to the usefulness of Fuller's extension. For example it could lead to a solution of the following question.

Suppose that G is a (compact) Lie group and that T^n is its maximal torus. Let

$$m : G \times T \to G$$

be the action of T^n on G by multiplication on the right. Let $\tau : \mathbb{R}^n \to T^n$ be the universal covering space of T^n, and let

$$m' : G \times \mathbb{R}^n \to G$$

be defined by $m'(x,v) = m(x,\tau(v))$.

Q. Is it the case that any $\mathbb{R}^n$ action on G sufficiently close to m' must have a closed torus?

This is one natural generalization of Seifert's classical result.

Bibliography

1. N. Bobylev and M. Krasnosel'skii, Operators with continua of fixed points, Dokl. Akad. Nauk, SSSR 205 (1972) (Soviet Math. Dokl. 13 (1972), 1045-1049).

2. R. Brown, An elementary proof of the uniqueness of the fixed point index, Pac. J. Math. 35 (1970), 549-558.

3. F. Fuller, An index of fixed point type for periodic orbits, Amer. J. Math., 89 (1967), 133-148.

4. R. Knill, On the homology of a fixed point set, Bull. Amer. Math. Soc., 77 (1971), 184-190.

5. J. Leray, Sur les equations et les transformations, J. Math. Pures Appl. (9) 24 (1945), 201-248.

6. H. Seifert, Closed integral curves in 3-space and isotopic two-dimensional deformations, Proc. Amer. Math. Soc., 1 (1950), 287-302.

7. A. Dold, Fixed point index and fixed point theorem for Euclidean neighborhood retracts, Topology 4 (1965), 1-8.

An application of the ρ-invariant

by

Friedrich Hegenbarth*

Introduction

The ρ-invariant is the difference of the two sides of the G-signature formula for G-actions on even-dimensional manifolds with boundary. It is an invariant of the G-action restricted to the boundary.

In [6] is given an expression of ρ for an S^1-action on spheres in terms of splitting invariants (Theorem 14 C.4). In this paper we will consider S^3-actions on spheres and express the restriction of ρ to S^1 by splitting invariants.

Let X^{4n} be a closed p.l.-manifold of dimension $4n$ which is homotopy equivalent to the quaternionic projective space HP^n and let $h: X \to HP^n$ be a homotopy equivalence. Assume that h is transversal-regular on $HP^k \subset HP^n$. We define the integer

$$a_k = 1/8 \ (\text{Index } (h^{-1}(HP^k)) - \text{Index } (HP^k))$$

(see e.g. [2] for a proof that a_k is an integer).

We pull the fibration $S^{4n+3} \to HP^n$ back to X and obtain a fibration $N^{4n+3} \to X$. N is p.l.-isomorphic to S^{4n+3} and therefore X is the orbitspace of an S^3-action on S^{4n+3}.

---

* Supported by NSF grant MPS 72-05055 A02

In §I we will define the ρ-invariant for this action and in §2 we will prove

$$(I.1) \qquad \rho(t) = f^{n+1} + \sum_{1\le r\le n-1} 8a_r(f^{n-r+1} - f^{n-r-1})$$

for $t \in S^1 - \{1\}$. Here f is the function $(1+t)^2/(1-t)^2$. The formula for the induced S^1-action is then (see [6] 14 C.4)

$$(I.2) \qquad \rho(t) = f^{2n+2} + \sum_{1\le r\le n} 8s_r\ (f^{2n-2r+2} - f^{2n-2r}),$$

where $f(t) = (1+t)/(1+t)$ and $t \in S^1 - \{1\}$.

The s_r are defined as follows: We have the S^2-fibration

$$CP^{2n+1} \to HP^n$$

which induces an S^2-fibration $V^{4n+2} \to X^{4n}$ via h. V is homotopy equivalent to CP^{2n+1}. We assume the homotopy equivalence $g\colon V \to CP^{2n+1}$ is transversal-regular on $CP^k \subset CP^{2n+1}$ and define

$$s_k = \frac{1}{8}\,(\text{Index}\ (g^{-1}(CP^{2k})) - \text{Index}\ (CP^{2k})).$$

Because I.1 and I.2 coincide on S^1-{1} one obtains necessary and sufficient conditions to extend an S^1-action to an S^3-action. This reproves a theorem of B. Conrad [3]. We also derive a formula for the splitting-invariants of a "join" action.

I would like to express my thanks to The Institute for Advanced Study for supporting me and S. Morita for many helpful discussions.

§I - Definition and properties of the ρ-invariant

ρ is defined in [1] (there the symbol σ is used). We will follow [4] in our definition.

Let M^{4n+4} be an orientable manifold with or without boundary of dimension $4n+4$ and let G be a compact group acting on it. Then the complex number

$$\mathrm{Sign}\ (g,M)$$

can be defined for every $g \in G$. In our case this number will be 0 or 1 depending on the dimension of M. Therefore we refer to [1] or [4] for a definition.

But we have to define the "other side" of the G-signature formula. Let M^g be the fixed point set of $g \in G$. We consider actions where M^g is an orientable manifold with a tubular neighborhood homeomorphic (equivariantly) to a G-vectorbundle N^g over M^g. Moreover we assume $M^g = M^{g'}$ for any $g' \in G - \{1\}$. Such an action is called tame and semifree.

Assume that the normal bundle N^g splits

$$(1) \qquad N^g = N^g_\pi \oplus \sum_{0<\theta<\pi} N^g_\theta ,$$

where N^g_π is the subbundle on which g acts by multiplication with -1 and N^g_θ is the subbundle on which g acts by direct sums of

$$A_\theta = \begin{pmatrix} \cos\theta & -\sin\theta \\ \sin\theta & \cos\theta \end{pmatrix}$$

The eigenvalue $+1$ can not occur . Because the representations given by A_θ and $A_{-\theta}$ are equivalent the summation in (1) is over all θ between 0 and π.

For any complex vectorbundle ξ of complex dimension r

let

$$L_\theta(\xi) = (\coth(i\theta/2))^r \prod_j \coth(x_j+i\theta/2)/\coth(i\theta/2).$$

The x_j are defined by the total Chern class of ξ

$$c(\xi) = \prod_j (1+x_j).$$

For any real vectorbundle ξ let

$$L(\xi) = \prod_j y_j/\tanh(y_j)$$

be the Hirzebruch L-class of ξ. The y_j are defined by the total Pontrjagin class

$$P(\xi) = \prod_j (1+y_j^2) .$$

Furthermore we define as in [4]

$$L_\pi(\xi) = e(\xi)\ L(\xi)^{-1},$$

where $e(\xi)$ is the Euler class of ξ, and

$$L'(g,M) = L(M^g)\ L_\pi(N^g)\cdot \prod_{0<\theta<\pi} L_\theta(N^g).$$

If M is closed the G-signature theorem states

(I.2) $$\operatorname{Sign}(g,M) = L'(g,M)[M^g]$$

Suppose now that G acts freely on Y^{4n+3} and that $k*(Y,G)$ bounds (W^{4n+4}, G), i.e. there is a G-action which restricts to the given G-action on boundary$(W) = k\cdot Y$. Then we define

(I.3) $$\rho(g) = \frac{1}{k}\,(L'(g,W)[W^g] - \operatorname{Sign}(g,W)).$$

It follows from (I.2) that ρ is well defined.

We consider now the S^3-action on S^{4n+3} associated to the homotopy equivalence $h\colon X^{4n} \to HP^n$. Let E be the total space of the associated D^4-bundle of $S^{4n+3} \to X^{4n}$. The action of S^3 extends to a semifree action on E with $E^g = X$.

We will compute ρ of that action. Because S^3 is connected we have

$$\mathrm{Sign}(g,E) = \mathrm{Sign}(E) = \begin{cases} 0 & \text{for } n \text{ even} \\ 1 & \text{for } n \text{ odd.} \end{cases}$$

Let us write for a moment $\rho = \rho_G$ to indicate that ρ depends on G. Then it is obvious that for any subgroup $H \subset G$

(I.4) $$\rho_G/H - \{1\} = \rho_H.$$

Let G act on S^{4n+3} and on S^{4m+3}. Then G acts via the diagonal on the join $S^{4n+3} * S^{4m+3} = S^{4(n+m+1)+3}$. Because the set of elements of finite order is dense in G the same proof as for 14 A.I in [6] gives

(I.5) The value of ρ for the join action is the pointwise product of the values of ρ for the given actions.

Furthermore we have

(I.6) If $g \in S^1 - \{1\} \subset S^3 - \{1\}$ then

$$\rho(g) = \frac{ge^{2x} + 1}{ge^{2x} - 1} \cdot \frac{\bar{g}e^{-2x} + 1}{\bar{g}e^{-2x} - 1} \cdot L(X^{4n})\,[X^{4n}] - \mathrm{Sign}(HP^n),$$

where $x^2 \in H^4(X^{4n})$ is a generator.

Proof: Let V^{4n+2} be the pull-back of the S^2-fibration $CP^{2n+1} \to HP^n$ and H the homotopy equivalence $V \to CP^{2n+1}$ covering h. Furthermore let $p\colon V \to X$ be the projection map. Then $p^*(E)$ contains canonically a complex line bundle η, which is invariant under the S^1-action. Choose a metric and let η' be the complement of η. We can choose complex structures on η and η' such that the eigenvalues are g (on η) and $\bar{g}$ (on η'). The corresponding representations are

$$\begin{pmatrix} \cos\theta & -\sin\theta \\ \sin\theta & \cos\theta \end{pmatrix} \quad \text{and} \quad \begin{pmatrix} \cos -\theta & -\sin -\theta \\ \sin -\theta & \cos -\theta \end{pmatrix} .$$

Therefore they appear with the same eigenvalue in (1) and hence in $L'(g,E)$. Because $p^*: H^*(X) \to H^*(V)$ is injective we may compute $p^*L'(g,E)$.

If $g \neq -1$ we have

$$p^*L'(g,E) = \coth(x+i\ \theta/2)\cdot\coth(x'+i\ \theta/2)\cdot L(X),$$

with $x = c_1(\eta)$, $x' = c_1(\eta')$.

Now $x + x' = p^*(c_1(E)) = 0$. This proves (I.6).

(I.7) Remark: Because in

$$(ge^{2x} + 1)(ge^{-2} + 1)$$

appear only squares of x and $p^*: H^{4*}(X) \to H^{4*}(V)$ is an isomorphism the above formula makes sense.

§2 - Proof of formula (I.I) and application of it.

We recall the diagram

$$\begin{array}{ccc} V^{4n+2} & \xrightarrow{H} & CP^{2n+1} \\ {\scriptstyle p}\downarrow & & \downarrow \\ X^{4n} & \xrightarrow{h} & HP^{n} \end{array}$$

where h is the given homotopy equivalence. In the Introduction we defined the integers s_k and a_k for $1 \leq k \leq n$. h determines an S^3-action on S^{4n+3}. For this action we have

Theorem 3.1 - If $t \in S^1 - \{1\}$ then

$$\rho(t) = f^{n+1} + \sum_{1 \le r \le n-1} 8a_r(f^{n-r+1} - f^{n-r-1}),$$

where $f(t) = (1+t)^2/(1-t)^2$.

Proof: The proof is similar to the proof of 14 C.4 [6].

From (I.6) follows

$$\rho(t) = \frac{te^{2x}+1}{te^{2x}-1} \cdot \frac{te^{-2x}+1}{te^{-2x}-1} L(X)[X] - \mathrm{Sign}(HP^n).$$

ρ is linear in the a_k, thus we may write

$$\rho = c_n + b_n^1 a_1 + \dots + b_n^{n-1} a_{n-1}.$$

For $n = 0$ we obtain $\rho(t) = (t+1)^2/(t-1)^2 = f(t)$.

Now we use the fact that ρ is multiplicative for join actions. We consider the join $S^{4n-1} * S^3$. For the S^3-action on S^{4n-1} we have

$$\rho = c_{n-1} + b_{n-1}^1 \cdot a_1 + b_{n-1}^2 \cdot a_2 + \dots + b_{n-1}^{n-2} \cdot a_{n-2}.$$

Therefore

$$b_n^r = b_{n-1}^r \cdot f \quad \text{and}$$

$$c_n = c_{n-1} f .$$

Iterating the process gives

$$b_n^r = b_{r+1}^r \, f^{n-r-1} \quad \text{and}$$

$$c_n = f^{n+1}.$$

It remains to compute b_n^{n-1}, for any n.

First we look for the coefficient of a_{n-1} in $L(X)$. Let this be $\alpha_{n-1} x^{2n-2}$. Let $W^{4n-4} \subset X^{4n}$ be a submanifold dual to $x^2 \in H^4(X)$. Then $\mathrm{Index}(W) = L(W)[W]$ by the index theorem. But $L(W) = L(\tau W)$ and $\tau W \oplus \nu W = i^* \tau X$, with $i: W \subset X$ the inclusion and

νW the normal bundle of i. Therefore $\text{Index}(W) = i^*(L(X)) \cdot L(\nu W)^{-1}[W]$. Because the constant term in $L(\nu W)^{-1}$ is 1, the coefficient of a_{n-1} in $\text{Index}(W)$ must be α_{n-1}. On the other hand

$$\text{Index}(W) = \text{Index}(HP^{n-1}) + 8a_{n-1}.$$

therefore we get $\alpha_{n-1} = 8$.

The coefficient of a_{n-1} in ρ therefore is

$$8 \cdot (\text{coefficient of } x^2 \text{in } \frac{te^{2x}+1}{te^{2x}-1} \frac{te^{-2x}+1}{te^{-2x}-1}).$$

It is now easy to see that coefficient is f^2-1.

This yields

$$b_n^r = 8f^{n-r-1}(f^2-1),$$

giving the above formula.

<u>Corollary 3.2</u> - (see [3]) - There are the following relations among the splitting invariants :

$$s_r = a_r + a_{r-1}.$$

Proof: (I.4) applies with $H = S^1 \subset S^3$. Comparing (I.I) and (I.2) yields

$$s_n(f^2-1) + s_{n-1}(f^4-f^2) + s_{n-2}(f^6-f^4) + \ldots =$$

$$a_{n-1}(f^4-1) + a_{n-2}(f^6-f^2) + a_{n-3}(f^8-f^4) + \ldots \quad .$$

Thus we obtain

1. $a_{n-1} = s_n$. Furthermore $s_n - s_{n-1} = -a_{n-2}$.

Using 1. we obtain

2. $a_{n-1} + a_{n-2} = s_{n-1}$.

The coefficient of f^4 gives $s_{n-1} - s_{n-2} = a_{n-1} - a_{n-3}$. Using 2. we have

3. $a_{n-2} + a_{n-3} = s_{n-2}$, etc... This proves 3.2.

Corollary 3.3 (See [3]. Thm. 1) - With the above notation

$$\sum_{1\le r\le n} (-1)^r s_r = 0.$$

Proof: We may apply 3.2 or compare (I.I) and (I.2) at $i = \sqrt{-1}$.

More generally, let $M^{4n} \to HP^n$ be a normal map of degree 1 corresponding to $f: HP^n \to G/PL$ and let $\sigma(f)$ be its surgery obstruction. As before let $E \to HP^n$ denote the associated D^4-bundle of $S^{4n+3} \to HP^n$. Then f extends to E and because $\pi_{4n+3}(G/PL) = 0$ also to a map $HP^{n+1} \xrightarrow{F} G/PL$. Applying the plumbing theorem we may assume that $\sigma(F) = 0$. Thus up to normal cobordism we can assume that

$$M^{4n} = H^{-1}(HP^n)$$

for some homotopy equivalence $H: X^{4n+4} \to HP^{n+1}$ (X^{4n+4} is a homotopy quaternionic projective space). Then we have

Corollary 3.4 - The surgery obstruction of the normal map $M^{4n} \to HP^n$ is (up to sign) $\sum_{1\le r\le n} (-1)^r s_r$.

Proof: We can apply 3.2. But now we have $s_n = a_n + a_{n-1}$ and $a_n = \sigma(F)$.

Corollary 3.5 - Let S^3 act on S^{4n+3} and on S^{4m+3} and let $a_1, a_2, \ldots, a_{n-1}$ and $b_1, b_2, \ldots, b_{m-1}$ be the associated splitting invariants. Then the splitting invariants of the join action $c_1, c_2, \ldots, c_{n+m}$ are given by the formula (with $k = n+m+1$)

$$c_r = a_r + b_r + 8 \sum_{i+j=r} a_i b_j + 8 \sum_{i+j=r-2} a_i b_j .$$

Proof: It follows from (I.r) that $f^{k+1} + 8\,\Sigma c_r f^{k-r-1}(f^2-1) =$

$$= (f^{n+1} + 8\,\Sigma a_i f^{n-i-1}(f^2-1))(f^{m+1} + 8\,\Sigma\, b_j f^{m-j-1}(f^2-1)).$$

Comparing the coefficients on both sides of $f^{k-r-1}(f^2-1)$ gives 3.5.

References

[1] Atiyah, M.F., Singer, I.M. - The index of elliptic operators III, Ann. of Math. 87 (1968), 546-604.

[2] Browder, W. - Surgery on simply connected manifolds, Ergebnisse Springer Verlag 1970.

[3] Conrad, B. - Extending free circle actions on spheres to S^3 actions. Proc. of the A.M.S., Vol. 27 (1971), 168-174.

[4] Hirzebruch, F., Zagier, D. - Application of the Atiyah-Bott-Singer theorem in topology and elementary number-theory, Publish or Perish, 1974.

[5] Sullivan, D. - Triangulating and smoothing homotopy equivalences and homeomorphisms,Geometric Topology seminar, Princeton University 1970.

[6] Wall, C.T.C. - Surgery on compact manifolds, Academic Press 1970.

G-Transversality to $\mathbb{C}P^n$

by Italo José Dejter *

1. Presentation and Results.

The following question arises from the work of T.Petrie (see [4], page 147). Let G be a compact Lie group, Y be a smooth G-manifold and F a G-fibre bundle map between G-vector bundles of the same dimension over Y which, on each fibre is proper and has degree one. Such a map is called a quasi-equivalence of G-vector bundles [1]. For examples, see section 4 .

<u>Question 1.1.</u> <u>What are necessary and sufficient conditions for the existence of a smooth</u> G-<u>map</u> $\bar{F}$ <u>properly</u> G-<u>homotopic to</u> F <u>and transverse to the zero section ?</u>

A related question that we will not discuss in this work is: When is $\bar{F}$ restricted to the inverse image of the zero section a homotopy equivalence ? This illustrates the interest of our answer to Question 1.1, namely, that the conclusion of Theorem 1.2 below could be used to construct exotic G-actions in the spirit of [4].

Let G be abelian,let Ω be a complex G-module and let $Y=P(\Omega)$ be the space of complex lines of Ω, which is naturally a G- manifold whose underlying manifold is a complex projective space. Then a quasi-equivalence F' of complex G-vector bundles over Y can be described as follows:

Let $G' = H\times G$, where $H = S^1$, and Ω' be a complex G'-module such that Ω' restricted to $\{1\}\times G$ is Ω and such that $H\times 1 \subset G'$

---

* Supported by a grant from Fundação de Amparo à Pesquisa do Estado de São Paulo.

acts freely on Ω'. Let $f':N'\longrightarrow M'$ be a quasi-equivalence of complex G'-modules, such that $N' \neq M'$. If we consider

$$\beta' = S(\Omega')\times_H M' \quad \text{and} \quad \gamma' = S(\Omega')\times_H N'$$

as G-vector bundles over Y, we define $F':\gamma'\longrightarrow\beta'$ by $F'[y;\zeta] = [y;f'(\zeta)]$, where $y \in S(\Omega')$ and $\zeta \in N'$.

We can always express f' as a direct sum $f' = f\oplus f_A$, where

(1.1a) $$f:N\longrightarrow M \quad \text{and} \quad f_A:A\longrightarrow A$$

are quasi-equivalences of complex G'-modules.

We can write $G = G_0\oplus G_1$, where G_0 is a finite group and G_1 is a connected group. Let $i:G_1\longrightarrow G'$ be the obvious inclusion and let

$$i^*M = \sum_{\chi\in\hat{G}_1} a_\chi\cdot\chi$$

where $\hat{G}_1$ is the set of irreducible real G_1-modules.

Theorem 1.2. *Assume that G_1 is infinite and that if $\chi \neq 1$ then $a_\chi \leqq 1$. A necessary, [resp. sufficient] condition for the existence of a smooth G-map $\bar{F}'$ properly G-homotopic to F' and transverse to the zero section is that for some choice of modules M and N as in (1.1a):*

a. H acts trivially on M and N, and

b.[resp. b'] There exists a complex G-module Γ such that Ω is a G-submodule of $\Lambda(M)\otimes\Gamma$ (where $\Lambda(M)$ is the total exterior power of M) containing at least one copy of each irreducible complex G-submodule of $\Lambda(M)\otimes\Gamma$ [resp., $\Omega = \Lambda(M)\otimes\Gamma$].

Part a. of Theorem 1.2 shows that the underlying vector bundles to β' and γ' over $\mathbb{C}P^n$ are trivial. This means that any exotic action of G on a homotopy $\mathbb{C}P^n$ constructed from β' and γ' via transversality would be tangentially equivalent to CP^n, supporting Petrie's conjecture on S^1- actions on homotopy $\mathbb{C}P^n$.

For the special case in which $\beta' = Y\times N'$, $\gamma' = Y\times M'$, we have that $F' = 1_Y\times f'$. In this case Theorem 1.2 reduces simply to conditions b. and b'. respectively, since condition a. is automatically satisfied.

Remark 1.3. The proof of the necessary, [resp. sufficient] condition of Theorem 1.2, to be given respectively in sections 2 and 3, factors through the fact that

i. For every $y \in Y^G$, β'_y is a real factor of

$$\nu_y(Y^G,Y) \oplus \gamma'_y$$

where $\nu_y(Y^G,Y)$ is the normal bundle of Y^G in Y at y. (This is essentially the method used in [4], page147).

ii. All the obstructions $O_*(K)$ to the existence of $\bar{F}'$ defined in Petrie G-transversality,[2] and [3],(where K are the isotropy subgroups of Y) must be defined, and the cohomology groups to which these obstructions belong must vanish for degrees bigger than 1, while $O_0(K) = 0$.

Remark 1.4. The sufficient condition of Theorem 1.2 is not necessary, as shown by Example 4.7. On the other hand, the difficulty in showing that the necessary condition of Theorem 1.2 is sufficient is that it does not necessarily imply ii. of Remark 1.3. as can be seen in Examples 4.9 and following.

Removing the assumption of Theorem 1.2 on G_1 and M creates a similar difficulty, see Examples 4.4 and 4.6.

2. Proof of the Necessary Condition.

We may suppose $G = S^1$. Let t^h denote the complex S^1-module given by $t\cdot z = t^h z$, where $t \in S^1$, $z \in C$ and $h \in \mathbf{Z}$. Let

$w = \max\{|h| \mid t^h$ is an irreducible complex S^1-submodule of $\Omega\}$.

Notation. We denote

(2.1) $\alpha(\bigoplus_{j=1}^{h} t^{a_j}) = \{a_j\}_{j=1}^{h}$, where $t \in S^1$ and a_j are integers.

We suppose $w = \max(\alpha(\Omega))$. By 1.3 i., this implies that if the necessary condition, a., of Theorem 1.2 were not true, then also

$$w_1 = w - r\, bw-a \quad \in \alpha(\Omega)$$

where $N(\eta^b t^a) \longrightarrow M(\eta^b t^a)$ is a summand component of the quasi-equivalence $f: N \longrightarrow M$, $a > 0$, $b > 0$ and

$r = \max\{|s| \mid t^s$ is an irreducible complex S^1-submodule of $M(t)$ not appearing in $N(t)\}$.

Of course, $w_1 < w$. Suppose

$$k = r|bw-a| = r(bw-a).$$

Then $w_2 = w_1 - r(bw_1 - a) \in \alpha(\Omega)$ necessarily, because

$$w_1 + r(bw_1-a) > w.$$

Observe $w_2 = w - kbr$. In the same way

$$w_3 = w_2 - r(bw_2 - a) \in \alpha(\Omega), \text{ and}$$

$$w_3 < w - kb^2r^2.$$

By induction we obtain $w_{n+1} < w - kb^n r^n$,

so that, for some n, $w_n < -w$, absurd.

The case $|bw-a| = a-bw$ could be treated nearly similarly. The proof works similarly if $-w = \max(\alpha(\Omega))$.

The rest of the statement, i.e. part b., follows from i. of Remark 1.3 and the assumption of Theorem 1.2 on M, by a similar observation as in [4], page 147.

3. Proof of the Sufficient Condition and Related Inequalities.

Proposition 3.1. Under the assumptions of Theorem 1.2, there exists a splitting

$$G = \bigoplus_{j=1}^{r+s} H_j,$$

where $H_j = S^1$ for $j = 1,\dots r$, H_j is finite for $j = r+1,\dots,r+s$ and such that the nontrivial complex irreducible summands of M are pairwise distinct as real H_j-modules for $j = 1,\dots,r$.

Proof. By the assumption of Theorem 1.2 on G_1, there always exists a splitting

$$G = \bigoplus_{j=1}^{r+s} H_{j1}$$

such that $H_{j1} = S^1$ for $j = 1,\dots,r$ and H_{j1} is finite for $j = r+1,\dots,r+s$.

The assumption of Theorem 1.2 on M allows us to assume that several nontrivial complex irreducible summands $M_1, M_2, \dots, M_{\ell_1}$ of M are pairwise distinct as real H_{11}-modules. If $\alpha(M|H_{j1}) = \{m_{jk}\}_k$, then we set $m_j = 1 + 2\max\{m_{jk}\}$. Then there is a splitting

$$G = \bigoplus_{j=1}^{r+s} H_{j2}$$

and a group isomorphism

$$\sigma: \bigoplus_{j=1}^{r+s} H_{j1} \longrightarrow \bigoplus_{j=1}^{r+s} H_{j2}$$

given by $\sigma(t_1,\dots,t_{r+s}) = (t_1',\dots,t_{r+s}')$, where $t_j \in H_{j1}$, $t_j' \in H_{j2}$ and $t_1' = t_1$, $t_j' = t_1^{m_j}.t_j$ for $j = 2,\dots,r$ and $t_j' = t_j$ for $j = r+1,\dots,r+s$. Moreover, $M_1,\dots,M_{\ell_1}$ are pairwise distinct as real H_{j2}-modules for $j = 1,\dots,r$. We can assume $\{M_1,\dots,M_{\ell_2}\}$ is the set of nontrivial irreducible complex summands of M which are pairwise distinct as real H_{22}-modules. If $\ell_1 = \ell_2$, the claim is true. If $\ell_1 < \ell_2$, we proceed similarly as above. Using induction, we obtain the desired splitting.

Notation. First, we set $\mathbb{Z}_0 = S^1$. To each isotropy subgroup $K \subset G$ of Y, there is associated an integer vector $\sigma = (\sigma_1, \ldots, \sigma_{r+s})$ in $\mathbb{Z}^{r+s}$ such that $K = \oplus_{j=1}^{r+s} K_j$, where $K_j \subset H_j$ and $K_j = \mathbb{Z}_{\sigma_j}$ for $j = 1, \ldots, r+s$. So we denote K by $\mathbb{Z}_\sigma$.

We set $t = (t_1, \ldots, t_{r+s}) \in G$, where $t_j \in H_j$. If $p = (p_1, \ldots, p_{r+s})$ in $\mathbb{Z}^{r+s}$, then we set $t^p = (t^{p_1}, \ldots, t^{p_{r+s}}) \in G$. Also we make use of notation (2.1) in the present context.

3.2. Proof of the sufficient condition of Theorem 1.2.

Under the assumptions of the sufficient condition of Theorem 1.2, there is a splitting $F' = F \oplus F_A$, where

$$F: \gamma \longrightarrow \beta \quad \text{and} \quad F_A: \phi_A \longrightarrow \phi_A$$

is a quasi-equivalence of complex G, (resp. G')-modules over Y, where

$$\gamma' = \gamma \oplus \phi_A \quad , \quad \beta' = \beta \oplus \phi_A$$

$$\gamma = Y \times N \quad , \quad \beta = Y \times M \quad , \quad \phi_A = S(\Omega') \times_H A \quad \text{and}$$

$F(y, \zeta) = (y, f(\zeta))$, $F_A[y, \zeta_A] = [y, \zeta_A(\zeta_A)]$, where $y \in Y$, $\zeta \in N$ and $\zeta_A \in A$.

Let K be an isotropy subgroup of Y and $\{Y_j^K\}$ the set of components of Y^G. Let

$$M = M^K \oplus M_K \qquad \text{and} \qquad N = N^K \oplus N_K$$

be the obvious splittings as K-modules. Let $\nu(Y_j^K, Y)_y$ be the G-normal bundle of Y_j^K in Y at $y \in Y_j^K$. Then there exist splittings of real K-modules

$$\nu(Y_j^K, Y)_y \oplus N_K = \sum_\chi a'_{j\chi} \cdot \chi \quad \text{and} \quad M_K = \sum_\chi a_{K\chi} \cdot \chi$$

where $\{\chi\}$ is the set of real K-modules which are restrictions of irreducible complex K-modules, i.e. $\{\chi\} = \{t^a$ as a real K-module; $a \in \mathbb{Z}^{r+s}\}$.

Then [3] establishes an obstruction

$$O_*(K) \in \prod_j H^*[Y_j^K/G,\ (Y_K \cap Y_j^K)/G,\ \pi_{*-1}(V'_{Kj})] = O^*(K)$$

to the existence of a smooth G-map $\bar{F}$ properly G-homotopic to F and transverse to the zero section, whenever $O_*(J)$ is defined and zero for every isotropy subgroup $J > K$ of Y. Here $Y_K = \bigcup_{J>K} Y^J$ and $V'_{Kj} = \prod_\chi V'^{\chi}_{Kj}$, where

$$(3.2a)\quad V'^{\chi}_{Kj} = \begin{cases} V^{\mathbb{R}}_{2a'_{j\chi},\,2a_{K\chi}}, & \text{if } K=\mathbb{Z}_\sigma,\ \sigma \text{ is even and } \chi=t^{\pm\sigma/2}. \\ V^{\mathbb{C}}_{a'_{j\chi},\,a_{K\chi}}, & \text{otherwise.} \end{cases}$$

($V^{\mathbb{K}}_{c,b}$ is the Stiefel manifold over $\mathbb{K}$ of b-frames in c-space [5]).

In our case, when $O_0(K)$ is defined, it is null. If X is a k-connected space, but $\pi_{k+1}(X) \neq 0$, we define Con(X) = k.

On the other hand

<u>Proposition 3.2b.</u> <u>For every isotropy subgroup</u> K <u>of</u> Y <u>and component</u> Y_j^K, <u>we have that</u> V'_{Kj} <u>is connected.</u>

(The proof of Proposition 3.2b is completed in 3.10).

Proposition 3.2b and [3] imply that $O_1(K)$ is defined and null for every isotropy subgroup K of Y.

<u>Proposition 3.3.</u> <u>For every isotropy subgroup</u> K <u>of</u> Y, <u>and component</u> Y_j^K <u>and Stiefel manifold</u> V'^{χ}_{Kj}, <u>we have that</u> $\dim(Y^K/G) \leq 1 + \mathrm{Con}(V'^{\chi}_{Kj})$, <u>for every</u> χ <u>such that</u> $a'_{K\chi} > 0$.

(The proof of Proposition 3.3 is completed in 3.11).

Proposition 3.3 implies that, if $a'_{Kj} > 0$, then

$$\dim(Y_j^K/G) \leq \min_{\{\chi;\, a_{K\chi}>0\}} [1 + \mathrm{Con}(V'^{\chi}_{Kj})] < 1 + \mathrm{Con}\, V'_{Kj}.$$

Then $O^{*+2}(K) = 0$. The same happens if there is no χ such that $a_K > 0$. This shows that $O^*(K) = 0$. Induction on the lattice of isotropy subgroups of Y completes the proof of the existence of a smooth G-map $\bar{F}$ properly G-homotopic to F and transverse to the zero section.

Now, condition a. of Theorem 1.2 can be used to insure validity of the proof above for the annihilation of the obstructions leading to the existence of the claimed map $\bar{F}'$.

A Combinatorial Setting and Proof of the Inequalities.

We proceed to establish the principles by means of which we prove Propositions 3.2b and 3.3. We can assume $M = \bigoplus_{i=1}^{n} t^{p_i}$ and $= t^0 \oplus (\bigoplus_{k=n+1}^{n+m} t^{p_i})$.

Let $D = \{-1,0,1\}$ and $\mathcal{D} \subset D^m$ be the subset of elements containing either at most one nonzero entry or exactly two nonzero entries opposite in sign, if $m \geq 2$; $\mathcal{D} = D$ if $m = 1$; $\mathcal{D} = \emptyset$ if $m = 0$. We define $d\colon D^n \times \mathcal{D} \longrightarrow \mathbb{Z}^{r+s}$ by $d(c_1,\ldots,c_{n+m}) = \sum_{i=1}^{n+m} c_i p_i$.

We say that $B \subset D^n \times \mathcal{D}$ is saturated iff, for every linear combination $b = \sum_{i=1}^{n+m} \gamma_i b_i \in D^n \times \mathcal{D}$ of elements $b_i \in B$ with coefficients $\gamma_i \in D$, we have $b \in B$.

We proceed to supply the central example of a saturated $B \subset D^n \times \mathcal{D}$ that we will need. We recall that $\mathbb{Z}_0 = S^1$. Let $\phi(Y)$ be the genus of Y, i.e. a maximal collection of pairwise distinct integer vectors σ such that $\mathbb{Z}_\sigma$ is an isotropy subgroup of Y. If $\sigma = (\sigma_1,\ldots,\sigma_{r+s}) \in \mathbb{Z}^{r+s}$ and $\tau = (\tau_1,\ldots,\tau_{r+s}) \in \mathbb{Z}^{r+s}$, then we say that τ is a multiple of σ iff τ_i is a multiple of σ_i, for $i = 1,\ldots,r+s$. We observe that the only integer multiple of 0 is 0. Then we define $\rho\colon \phi(Y) \longrightarrow \mathcal{P}[\phi(Y)]$ by $\rho(\sigma) = \{\tau \subset \phi(Y)\colon \tau$ is a multiple of $\sigma\}$.

If $I \subset \mathbb{Z}^{r+s}$, let $\zeta(I) = \{\pm\sigma;\ \sigma \in I\}$.

Let $K = \mathbb{Z}_\sigma$ be an isotropy subgroup of Y. Then

<u>Proposition 3.4.</u> $B = d^{-1}\zeta\rho\,(\sigma)$ <u>is saturated in</u> $D^n \times \mathcal{D}$.

Proof. d is a surjection preserving linear combinations of elements of $D^n \times \mathcal{D}$ with coefficients in D and values in $D^n \times \mathcal{D}$ Let $b_i \in B$ for $i = 1,\ldots,r$. Then $d(b_i) = k_i\sigma$, where $k_i \in \mathbf{Z}$. Let $b = \sum_{i=1}^{r} \gamma_i b_i \in D^n \times \mathcal{D}$, where $\gamma_i \in D$. Then

$$d(b) = \sum_{i=1}^{r} \gamma_i d(b_i) = \sigma\left(\sum_{i=1}^{r} \gamma_i k_i\right) \in B, \text{ as claimed.}$$

Let $E = \{0,1\}$ and $\mathcal{E} = E^m \cap \mathcal{D}$. Two elements x and y of $E^n \times \mathcal{E}$ are said to be B-equivalent iff $x-y \in B$.

Let $\{U_j\}$ be the set of B-classes of $E^n \times \mathcal{E}$. The number of elements of a finite set Q will be denoted $o(Q)$.

Proposition 3.5. We can assume that the subindexing of the Y_j^K's and the U_j's is such that $Y_j^K = \mathbb{C}P^{o(U_j)-1}$.

Proof. Let $S = \bigoplus_{y \in Y'} \tau_G(Y)_y$, where $Y' = \{y \in Y$: exactly one homogeneous coordinate of y is nonzero$\}$ and $\tau_G(Y)$ is the G-tangent bundle of Y at $y \in Y'$. Then there exists a bijection $\psi : D^n \times \mathcal{D} \longrightarrow \alpha(S)$ such that:

i. $d = e \circ \psi$, where $e : \alpha(S) \longrightarrow \mathbb{Z}^{r+s}$ is the obvious numerical evaluation;

ii. ψ preserves linear combinations in $D^n \times \mathcal{D}$ with coefficients in D and values in $D^n \times \mathcal{D}$, and

iii. ψ maps $E^n \times \mathcal{E}$ onto $\alpha(\Omega)$.

We define $a \equiv b \pmod{0}$ iff $a-b$ is a multiple of 0, that is iff $e(a) = e(b)$. Then we observe that $\psi\{(U_j)\}$ is the collection of classes modulo σ of $\alpha(\Omega)$ and that we can assume

$$Y_j^K = P[\alpha^{-1}\psi(U_j)] = \mathbb{C}P^{o(U_j)-1}.$$

Let $\Delta \subset E^n$ be the subset of elements having at most one nonzero entry. Two elements x and y of $\Delta \times 0 \subset \mathbb{Z}^{n+m}$ are said to be up-to-sign B-equivalent iff x is B-equivalent to either y or $(-y)$.

Let $\mathcal{B}_0 = \{Q_\ell\}$ be the set of up-to-sign B-classes of $\Delta\times 0$ and $\mathcal{B} = \mathcal{B}_0 - \{Q_0 = B\cap(\Delta\times 0)\}$.

The K-normal bundle of Y_j^K in Y at a point $y \in Y^K$ admits a decomposition $\nu_K(Y_j^K,Y)_y = \oplus\, a_{j\chi}\cdot\chi$. Let $V_{Kj} = \prod_\chi V_{Kj}^\chi$, where $V_{Kj}^\chi = V^{\mathbb{C}}_{a_{j\chi},a_{K\chi}}$.

If $a \in \mathbb{Z}^\ell$, we define $\theta_a:\mathbb{Z}^\ell \longrightarrow \mathbb{Z}^\ell$ by $\theta_a(b) = b+a$.

Proposition 3.6. We have that

$$V_{Kj} = \begin{cases} \prod_\ell V^{\mathbb{C}}_{\mu_{j,\ell},o(Q_\ell)}, & \text{if } \mathcal{B}\neq\emptyset \text{ and } Q_\ell\in\mathcal{B}. \\ \text{point} & , \text{ if } \mathcal{B}=\emptyset \end{cases}$$

where $\mu_{j,\ell} = o\{[\theta_{-u_j}(E^n\times E)-B]\cap\bigcup_{\varepsilon=-1,1}\theta_{\varepsilon q_\ell}(B)\}$ is independent of $u_j \in U_j$ and $q_\ell \in Q_\ell$.

Proof. For each complex irreducible K-module χ such that $a_{K\chi} > 0$, there exists $c \in \mathbb{Z}^{r+s}$ such that χ is the K-restriction of the complex G-module t^c. Then, the set of elements in $\psi(\Delta\times 0)$ that are congruent modulo σ to either c or (-c) has order $a_{K\chi}$. In other words, $\psi^{-1}(A_c)\subset\Delta\times 0$ is an up-to-sign B-class of $\Delta\times 0$, and there is a labeling c_ℓ of those c not divisible by σ, such that ψ induces a bijection $\{Q_\ell\}\longrightarrow A_{c_\ell}$: Thus, if $\chi_\ell = t^{c_\ell}$, then $a_{K\chi_\ell} = o(Q_\ell)$.

Finally, we will show that $a_{j\chi_\ell} = \mu_{j,\ell}$. Let $y = P[\alpha^{-1}\psi(u_j)] \in Y_j^K$. Then

$$\alpha[\tau_G(Y_j^K)_y] = \theta_{-\psi(u_j)}[\psi(U_j)] \qquad \text{and}$$

$$\alpha[\tau_G(Y)_y] = \theta_{-\psi(u_j)}[\psi(\Omega)]$$

imply $\alpha[\nu_G(Y_j^K,Y)] = \theta_{-\psi(u_j)}[\alpha(\Omega)-\alpha(U_j)]$.

Then $a_{j\chi_\ell}$ is the order of the subset in $\theta_{-\psi(u_j)}[\alpha(\Omega)-\psi(U_j)]$

of elements congruent modulo σ to either c_ℓ or $(-c_\ell)$. Now, ψ maps $\theta_{-u_j}(E^n\times E)$ onto $\alpha[\tau_G(Y)_y]$, so that

$$a_{j\chi_\ell} = o\{\theta_{-u_j}(E^n\times E - U_j) \cap \bigcup_{\varepsilon=-1,1} \theta_{\varepsilon q_\ell}(B)\} = \mu_{j,\ell},$$

where $q_\ell \in Q_\ell$ is arbitrary. This proves the Proposition.

We define

$$\delta(\ell) = \begin{cases} 1, & \text{if } o(Q_\ell) < 1 \\ 0, & \text{if } o(Q_\ell) = 1 \end{cases}$$

and $\delta'(\ell) = \dim(Y^K) - \dim(Y^K/G)$.

Then the assumption of Theorem 1.2 on G_1, and M as modified in Proposition 3.1, implies

$$(3.7) \qquad \delta'(\sigma) \geq \delta(\ell).$$

<u>Proposition 3.8.</u> <u>Let</u> $B \subset D^n\times D$ <u>be saturated and</u> U_j <u>be a</u> B-<u>class in</u> $E^n\times E$, <u>and</u> $Q_\ell \in B$. <u>Then</u>

$$o(U_j) - 1 - \delta(\ell) \leq \mu_{j,\ell} - o(Q_\ell).$$

<u>Proof.</u> Let $\Delta_i \in E^n\times E$ be the element with i^{th} coordinate equal to 1 and any other coordinate equal to 0. For convenience of notation, if $z \in D^h$, let $z = (z_1,\ldots,z_h)$ be the vector form of z and let us denote $z' = (z_1,\ldots,z_{h-1}) \in D^{h-1}$; if $A \subset D^n$, let $A' = \{z' \in D^{n-1};\ z \in A\}$; if $\xi = o(A)$, let $\xi' = o(A')$. (This notation will be used regardless of the symbols used in place of z, A or ξ to represent an element, a subset or an order of D^h).

Let $u \in E^n\times E$. Assume $\Delta_1 \notin B$ and let

$$\xi = o\{b\in B:\ b\in\theta_{-u}(E^n\times E)\} \quad ; \quad \eta = o\{b\in B:\ b\pm\Delta_1\in\Delta\}$$

$$\text{and} \quad \zeta = o\{b\in B:\ b\pm\Delta_1\in\theta_{-u}(E^n\times E)-B\}.$$

First we prove $\xi \leq \zeta$ and $\xi + \eta \leq \zeta$. We proceed by induction. The first step, n+m = 1 and n =1, is elementary. The inductive $(n+m)^{th}$ step for n > 1 has two cases: (where r = n+m)

<u>i.</u> $\Delta_r \in B$. Then direct inspection shows that $\xi = 2\xi'$, $\eta = \eta'$ and $\zeta \geq 2\zeta'$. So it is enough to see that

$$2\xi'+\eta'-2 \leq 2(\xi'+\eta'-2),$$

which is good for $\eta' \geq 2$. Notice that $2(\xi'+\eta'-2) \leq 2\zeta'$ by the inductive assumption. The case $\eta' \leq 1$ is elementary.

<u>ii.</u> $\Delta_r \notin B$. Then $\xi = \xi'$, and if we denote

$$\kappa = \begin{cases} o\{\Delta_1 \pm \Delta_r \subset B\}, & \text{if } \Delta_r \in \Delta\times 0 \\ 0 & , \text{if } \Delta_r \notin \Delta\times 0, \end{cases}$$

then $\eta = \eta'+\kappa$ and $\zeta = \zeta'+\kappa$.

So this proves our claims, which, in the notation of the statement of this proposition, can be stated

$$o(U_j) \leq \mu_{j,\ell} \quad \text{and} \quad o(U_j)-2 \leq \mu_{j,\ell} - o(Q_\ell)$$

which implies the proposition.

<u>Using the Fact that</u> F <u>is a Nontrivial Quasi-equivalence.</u>

We make $\alpha(M)$ into a lattice by means of the following rule

$$a < b \quad \text{iff} \quad a|b$$

If $a = (a_1,\ldots,a_{r+s}) \in \mathbb{Z}^{r+s}$, we set $|a| = (|a_1|,\ldots,|a_{r+s}|)$.

Let $\alpha'(M,N) =$

$\{a\in\alpha(M)$: a is maximal in $\alpha(M)$; $a \neq b$ for every $b\in\alpha(N)\}$.

The assumption of Theorem 1.2 on G_1 and M implies that the elements of $\alpha'(M,N)$ are pairwise distinct in absolute value; so we can assume $\alpha'(M,N) \subset \mathbb{Z}^{r+s}$. Let

$$\varepsilon(\sigma,\ell) = \begin{cases} 0, & \text{if } \sigma/2 \in \mathbb{Z}^{r+s} \text{ and } \alpha^{-1}d(Q_\ell) \equiv \sigma/2 \pmod{\sigma} \\ 1, & \text{otherwise.} \end{cases}$$

Let $\phi(\sigma,\ell) = o\{h \in \alpha(N) : h \equiv d(Q_\ell) \pmod{\sigma}$.

Proposition 3.9: If $\varepsilon(\sigma,\ell) = 0$, then

a. $d(Q_\ell) \subset \alpha'(M,N)$ implies $\delta(\ell) = 0$;

b. Otherwise, $\phi(\sigma,\ell) \geq 1$.

Proof. a. This is insured by the assumption of Theorem 1.2 on G_1 and M.

b. For each $a \in \alpha(M)$ which is neither maximal in $\alpha(M)$ nor an integer of $\alpha(N)$, there exists an element c, maximal in $\alpha(M)$, such that a divides two elements b_1 and b_2 and such that $|b_1| \neq |b_2|$, $b_1|c$, $b_2|c$ and $(b_1,b_2) = a$. So that we can assume $b_1 = (2k+1)a$, and so, if a is an integer vector in $d(Q_\ell)$, we have $b_1 \equiv a \pmod{2\sigma}$.

3.10. Proof of Proposition 3.2b. Proposition 3.9 implies that, if V'_{Kj} is not connected, then, for some $Q_\ell \in \mathcal{B}$, $\varepsilon(\sigma,\ell) = 0$ and $d(Q_\ell) \subset \alpha'(M,N)$. There exists a pair $\{a_1,a_2\} \subset \alpha(\Omega)$ such that $a_1 \equiv a_2 \pmod{\sigma}$. Then $a_i \in \alpha(M) \otimes \Gamma_i$, for $i = 1,2$, where Γ_1 and Γ_2 are irreducible complex G-summands of Ω such that $\Gamma_1 \neq \Gamma_2$. If $b = \psi(q_\ell)$, for some $q_\ell \in Q_\ell$, then there exists $\varepsilon_1,\varepsilon_2 = \pm 1$ such that $a_i + \varepsilon_i b$ is in $\alpha(\Omega)$ with the multiplicity of b in M_K, for $i = 1,2$.This shows that V'_{Kj} cannot be nonconnected.

<u>3.11. Proof of Proposition 3.3.</u> First we consider the case under the assumptions of Proposition 3.9a. By Proposition 3.5

$$\dim (Y_j^K/G) \leq 2\, o(U_j) - 3 ,$$

which by Proposition 3,9a and Proposition 3.8 is less or equal than $2[\mu_{j,\ell} - o(Q_\ell)]$, which, in turn, by Proposition 3.6, (3.2a) and [5], page 34 is less or equal than $1 + \mathrm{Con}(V_{Kj}'^{\,\chi_\ell})$.

For any other situation than the one given in Proposition 3.9a, we have by Proposition 3.5

$$\dim(Y_j^K/G) \leq 2\, o(U_j)-2-\delta'(\sigma),$$

which by (3.7) is less or equal than $2\, o(U_j)-2-\delta(\ell)$ when $d^{-1}[\alpha(\chi)] \in Q_\ell$, which, in turn, by Proposition 3.8 is less or equal than $[2\mu_{j,\ell}-o(Q_\ell)]+1$, which by Proposition 3.9b is less or equal than $2[\mu_{j,\ell} - o(Q_\ell)]+\phi(\sigma,\ell)+\varepsilon(\sigma,\ell)$, which, in turn, by Proposition 3.6, (3.2a) and [5], page 34 is less or equal than $1 + \mathrm{Con}\, V_{Kj}'$, either if $\varepsilon(\sigma,\ell) = 1$ or 0.

<u>4.1. Examples.</u>

We start by stating a version of Petrie-Meyerhoff Theorem[1] which will supply the material for the examples of the section. Let G be a compact abelian Lie group, R(G) the representation ring of G, $\dim: R(G) \longrightarrow \mathbb{Z}$ the dimension homomorphism given by $\dim(M-N) = \dim M - \dim N$, and ψ^k the kth Adams operation on R(G). Let $P = \{p_1,\dots,p_r\}$ be a collection of pairwise coprime positive integers, where $r > 1$, and $P'(t) = \Pi_{j=1}^{r} (\psi^{p_j} - 1)(t) \in R(G)$, $t \in G$. If $x \in \mathrm{Ker}(\dim) \subset R(G)$, we define $x \geq 0$ iff $x = M-N$ and there exists a quasi-equivalence $N \longrightarrow M$. Then

Theorem. Let $x \in \mathrm{Ker}(\dim) \subset R(G)$. Then $x \geq 0$ iff

$$x = \sum_{\chi} \sum_{P} a_{\chi,P} \cdot P'(\chi)$$

where $a_{\chi,P}$ are nonnegative integers , the sum is finite and χ ranges over the irreducible complex G-modules.

We remark that for every P as above P' is the difference of two complex G-modules, $M_P - N_P$, such that $r(P)=\lambda_{-1}(M_P)/\lambda_{-1}(N_P)$ is the product of cyclotomic polynomials. If P is a collection of primes, r(P) is a cyclotomic polynomial.

Examples. 4.1. Let $G=S^1$, $N=t^6+t^{10}+t^{15}+t^{1}$, $M=t^2+t^3+t^5+t^{30}$ and $\Omega = \Lambda(M)$. (We can associate the modules M and N with the collection $P=\{2,3,5\}$, as indicated above).

Let $f:N \longrightarrow M$ be a quasi-equivalence of complex G-modules, as guaranteed by the observations above, and define $F:P(\Omega)\times N \longrightarrow P(\Omega)\times M$ by $F(y,\zeta)=(y,f(\zeta))$. Then [2] implies the existence of a smooth G-map $\bar{F}$ properly G-homotopic to F and transverse to the zero section. In this particular example, the inequality of Proposition 3.3 becomes an equality at $y=[1;0;\dots;0]$ for $K=Z_2$. This could be used to infer that Proposition 3.3 does not hold for G finite.

4.2. Let $G=S^1$, $N=t^5+t^7+t^4+t^6$,and $M=t^1+t^{35}+t^2+t^{12}$. Then proceed as in the previous example, with a similar conclusion.

4.3. Let N and M be as in example 4.1. Let (see notation of section 1), $A=\eta^a t^b$, $\Omega'=\Lambda(M \oplus t^b)$, $N'= N\oplus A$ and $M'=M\oplus A$. Let $f'=f\oplus \mathrm{id}_A$ and $F':S(\Omega')\times_H N' \longrightarrow S(\Omega')\times_H M'$ be defined by $F'[y;\zeta]=[y;f'(\zeta)]$. Then Petrie G-transversality implies the existence of a smooth G-map $\bar{F}'$ properly G-homotopic to F' and transverse to the zero section.

4.4. The trouble with removing the assumption on M in the statement of Theorem 1.2 is that for $K=G=S^1$, ii. of Remark 1.3 is not necessarily satisfied as can be seen when $\alpha(N)=\{2,2,3,3\}$, $\alpha(M)=\{1,1,6,6\}$ and $\Omega = \Lambda(M)$.(see notation in section 2).

4.5. On the other hand, the module M may include any number of G-trivial components. For example, let $G=S^1$, $N=r.t^0+t^2+t^3$ and $M=r.t^0+t^1+t^6$. Let $\Omega=\Lambda(M)$, $Y=P(\Omega)$ and $f:N\longrightarrow M$ be a quasi-equivalence. Then there exists a smooth G-map $\bar{F}$ properly G-homotopic to $id_Y\times f$ and transverse to the zero section.

4.6. The following shows the necessity of the hypothesis of Theorem 1.2 that the complex irreducible summands of M must be pairwise distinct as real G-modules. Let $G=S^1$ and let

$\alpha(M)=\{-1,1,-6,6\}$ and $\alpha(N)=\{-2,2,-3,3\}$.

(Or $(M)=\{-1,1,-20,21\}$ and $\alpha(N)=\{-4,-5,3,7\}$).

Then i. of Remark 1.3 is not satisfied when taking Ω as the complex S^1-submodule of $\Lambda(M)$ with no repetition of complex irreducible summands.

The initial conjecture of T.Petrie, as suggested in [4], page 147, was that a necessary and sufficient condition for the existence of a smooth G-map F' properly G-homotopic to a quasi-equivalence $F:Y\times N\longrightarrow Y\times M$ is that $\Omega = \Lambda(M)\otimes\Gamma$ for some complex G-module Γ. The following is a counterexample to that conjecture.

4.7. Let $G=S^1$ and let

$N=t^2+t^3$, $M=t^1+t^6$ and $\Omega=1+t+t^2+t^5+t^6+t^7$.

If $f:N\longrightarrow M$ is a quasi-equivalence, define $F:P(\Omega)\times N\longrightarrow P(\Omega)\times M$ by $F(y,\zeta) = (y,f(\zeta))$. Then Petrie G-transversality [3] shows that there exists a smooth G-map $\bar{F}$ properly G-homotopic to F and transverse to $P(\Omega)\times 0$.

4.8. Let G,N.M be as above, and let $\Omega = 1+t+t^6+t^7+t^{12}+t^{13}$. In this case $O^1(K)$ is not defined because the corresponding Stiefel manifold is not connected. So a group of coefficients is not available. However, according to T. Petrie, the G-transversality of [3] could have been demonstrated using orientability of the bundles β' and γ' of section 1 in order to insure connectedness of the associated Stiefel manifold V'_{Kj}. The special case of $V'_{Kj} = Gl(k)/Gl(k-k')$ would reduce to $V'_{Kj} = SGl(k)/SGl(k-k')$. If $k = k'$, the troublesome nonconnectedness would disappear, insuring the definition and the annihilation of the obstructions $O_*(K)$.

The necessary condition of Theorem 1.2 does not necessarily imply ii. of Remark 1.3, as can be seen in the following examples.

4.9. Let $G=S^1$ and let N and M be as in Example 4.1, but take Ω as the submodule of $\Lambda(M)$ which has only one copy of each irreducible complex module. Let $K=\mathbb{Z}_2$. At the component Y_j^K which contains the point $y=[1;0;\ldots;0] \in P(\Omega)$, $H^*(Y_j^K/G;Y_j^K \cap Y_K/G)$ has top nonzero dimension equal to 13, while $\pi_*(V_{Kj})$ has bottom nonzero dimension equal to 12.

4.10. In a similar way, let $G=S^1$, $M=t+t^6$; $N=t^2+t^3$, $\Gamma = \Lambda(M)$ and $\Omega \subset \Lambda(M) \otimes \Gamma$ be the complex G-module having exactly one copy of each irreducible G-submodule of $\Lambda(M)\otimes\Gamma$. Then, if $O_*(\mathbb{Z}_2)$ were defined, it would be nonzero at $y=[1;0;\ldots;0]$.

4.11. Let $G=S^1$, $\Gamma = t^0+t^2$, $M=t^2+t^3+t^4+t^5+t^{60}$, $N=t^2+t^{12}+t^{15}+t^{20}+t$, and let Ω be such that $\Omega\oplus L=\Lambda(M)\otimes\Gamma$, where $L|\mathbb{Z}_4=8t^2$. Then, if $O^*(\mathbb{Z}_4)$ is defined, it is nonzero at $y=[1;0;\ldots;0]$.

Remark. On the last three examples we can select instead a different module Ω contained strictly in $\Lambda(M)\otimes\Gamma$, so as to have a smooth G-map $\bar{F}$ properly G-homotopic to F and transverse to $Y\times 0$.

REFERENCES

1. A. Meyerhoff and T. Petrie, Quasi-equivalences of G-modules, Topology, Volume 15, nº 1, 1976.

2. T. Petrie, G-Transversality, Bull. AMS, 81, nº 4, 1975.

3. ________, G-transversality, Aarhus Universitet Preprint Series 1975/76, nº 20, April 1976.

4. ________, Real algebraic actions on projective spaces, a survey, Annales de L'Institut Fourier, Tome XXIII, Fasc. 2, 1973.

5. N. Steenrod, The Topology of fibre bundles, Princeton University Press, 1951.

INSTITUTO DE CIÊNCIAS MATEMÁTICAS DE SÃO CARLOS, UNIVERSIDADE DE SÃO PAULO, 13.560-SÃO CARLOS-(SP), BRAZIL.

SOME PROBLEMS IN FOLIATION THEORY AND RELATED AREAS

Edited by Paul A. Schweitzer

This problem list, which originated in two problem sessions during the Symposium, has grown and been somewhat refined through the contributions and assistance of many mathematicians, to whom the editor expresses his gratitude. Special thanks are due to Michel Herman and James Heitsch. Whatever errors, omissions, or failures to give due credit may have persisted, despite the editor's efforts, are his exclusive responsibility, for which he asks the reader's kind indulgence.

For convenience, the problems have been grouped under the following rough headings:

1. Gelfand-Fuks cohomology, $B\overline{Diff}$, and characteristic classes of foliations.
2. Variation of foliations and stability.
3. Qualitative properties of foliations.
4. Minimal sets.

The interest aroused by a preliminary version of this problem list contributed to the solution of three of the original problems, which have consequently been deleted. May the present list, too, stimulate fruitful and enjoyable mathematical inquiry!

1. Gelfand-Fuks Cohomology, $B\overline{Diff}$, and Characteristic Classes of Foliations

1. Relate $H^*_{GF}(M)$ to $H^*(B\overline{Diff}\ M)$.

Here the Gelfand-Fuks cohomology of the smooth manifold M is denoted interchangeably by $H^*_{GF}(M)$ or $H^*(v_M)$, where v_M is the Lie algebra of smooth vector fields on M with the C^∞ topology, and the cohomology is the Lie algebra cohomology using only the continuous alternating multilinear forms on v_M as cochains. $\overline{Diff}\ M$ denotes a topological group, the homotopy fiber of the canonical homomorphism $Diff_\delta M \to Diff\ M$, where $Diff\ M$ is the group of C^∞ diffeomorphisms of M endowed with the C^∞ topology, and $Diff_\delta M$ is the same group with the discrete topology.

According to the folklore, if M is compact then $H^*_{GF}(M)$ should be isomorphic to the "differential cohomology" $H^*_d(B\overline{Diff}\ M)$ obtained using suitably defined "differentiable" cochains. Whether or not this is so, $H^*_d(B\overline{Diff}\ M)$ is a reasonable candidate for an intermediate link between $H^*_{GF}(M)$ and $H^*(B\overline{Diff}\ M)$. The analogy between the relationship of a Lie group to its Lie algebra, and the relationship of Diff M to v_M, is suggestive, as is the Van Est Theorem.

Ref. A. Haefliger, Cohomology of Lie algebras and foliations, these Proceedings.
R. Bott, Some remarks on continuous cohomology, Manifolds--Tokyo--1973, 161-170.

1.1. Conjecture. The universal homomorphism $H^*(v^c_{R^n}) \to H^*(B\overline{Diff}_K R^n)$ is injective. (Here c and K denote "compact support".)

An obvious method is to construct examples of trivialized foliated R^n bundles with compact support (i.e., outside a compact neighborhood of the zero section the foliation agrees with the trivialization), such that a given class in $H^*(v^c_{R^n})$ is mapped non-trivially by the characteristic homomorphism $H^*(v^c_{R^n}) \to H^*(X;R)$ into the cohomology of the base space X, assumed to be compact. A similar attack may be tried for Problem 1: construct foliations of $M\times X$ transverse to the first factor.

1.2. Show that if $H^i_d(B\overline{Diff}\ M) = 0$ for $i\le q$ (using an appropriate definition of differential cohomology), then $H^i(B\overline{Diff}\ M) = 0$ for $i\le q$.

1.3. Prove that $H^2(B\overline{Diff}_K R^2) = 0$.

This is the first unknown case of Thurston's conjecture that $H_k(B\overline{Diff}_K R^n) = 0$ for $k\le n$. Thurston has shown that the three groups $H_k(B\overline{Diff}_K R^n;Z)$, $H_k(B\overline{Diff}\ M^n;Z)$, and $H_{k+n}(B\overline{\Gamma}_n;Z)$ are isomorphic for the lowest dimension k for which one of them is non-zero. (Notation: $B\Gamma_n$ is the Haefliger classifying space for smooth codimension n foliations, and $B\overline{\Gamma}_n = F\Gamma_n$ is the homotopy fiber of the classifying map of the normal bundle $B\Gamma_n \to BO_n$.)

Ref. W. Thurston, Foliations and groups of diffeomorphisms, Bull. Amer. Math. Soc. 80 (1974) 304-307.

2. (Haefliger) Construct Γ_n-structures on S^{2n+1} with trivial normal bundles detecting elements of $H^{2n+1}(v^c_{R^n})$.

If $B\overline{\Gamma}_n$ is 2n-connected (a condition equivalent to Thurston's conjecture, see preceding problem), then by the Hurewicz Theorem all of $H^{2n+1}(B\overline{\Gamma}_n)$ is detected by maps $S^{2n+1} \to B\overline{\Gamma}_n$, so such Γ_n structures should exist. On $S^{8\ell+1}$ the Whitehead product $[\gamma,\gamma]$ lifted to $B\overline{\Gamma}_{4\ell}$, where $\gamma : S^{4\ell} \to B\Gamma_{4\ell}$ detects p_ℓ, is an example [P.Schweitzer and A. Whitman, Pontryagin polynomial residues of isolated foliation singularities, these Proceedings]. Many examples on other manifolds are known (Thurston, Heitsch, and others).

2.1. Calculate $H^*(B\Gamma_n)$.

3. (Heitsch) Relate $\pi_1 X$ to secondary characteristic classes, where X is the base of a foliated bundle.

<u>Conjecture</u> (Shulman). A foliated S^n bundle over the n+1 torus given by a representation $\pi_1 T^{n+1} \to \text{Diff } S^n$ has its characteristic homomorphism $H^*(\mathfrak{a}_n, O_n) \to H^*(T^{n+1})$ equal to zero. (See, for example, [R. Bott, On some formulas for the characteristic classes of group actions, these Proceedings] for the notation .)

M. Herman [The Godbillon-Vey invariant of foliations by planes of T^3, Geometry and Topology, Springer LNM 597 (1977),294-307] has shown for the case n=1 when the total space is T^3 and all the leaves are planes that the Godbillon-Vey invariant is zero. The conjecture also holds for many other examples (Heitsch).

4. (Haefliger) Given $k>1$, does there exist a manifold M and a representation $\pi_1 M \to \text{Diff } S^1$ such that the associated bundle has $\chi^k \neq 0$?

Such examples would give evidence for the plausible conjecture that the homomorphism $H^*(v_{S^1}, SO_2) \to H^*(B\text{Diff } S^1)$ is injective, since it is known that $\chi^k \neq 0 \in H^*(v_{S^1}, SO_2)$ for all k.

For the case k=1, Benzérci [Sur les variétés localement affines et localement projectives, Bull. Soc. Math. France 88(1960) 229-332] and Milnor [On the existence of a connection with curvature zero, Comment. Math. Helv. 32(1958),215-223] give examples when M is a surface.

5. (Heitsch) Let M^{n+1} be an orientable manifold and let $SL(M^{n+1})$ =

sup $|\chi(P)|$, where P varies over all S^n bundles over M^{n+1} with discrete structural group PSL_{n+1}.

Theorem (D. Sullivan) $SL(M^{n+1}) < \infty$. (In fact, $SL(M^{n+1})$ is less than or equal to the minimum number of $n+1$ simplexes in a simplicial triangulation of M^{n+1}.)

Problem. Determine the exact value of $SL(M^{n+1})$, perhaps in terms of known invariants of M^{n+1} such as the Pontryagin and Euler classes.

When $n=1$ and M is orientable, $SL(M^2) = \text{genus}(M^2)$ (Milnor, Benzẽrci, Wood).

Ref. D. Sullivan, A generalization of Milnor's inequality concerning affine foliations and affine manifolds, Comment.Math.Helv. 51 (1976),183-189.
J. Wood, Bundles with totally disconnected structure group, Comment.Math.Helv. 46 (1971),257-273.

6. (Haefliger) Let $V_{R^n}^{(k)}$ be the Lie algebra of k-jets in t of 1-parameter families of vector fields on R^n,

$$X = \sum_{i=0}^{k} t^i X_i \ , \quad X_i \in V_{R^n}, \quad t \in R, \quad 0<k\leq\infty \ ,$$

with the Lie bracket $[\Sigma t^i X_i, \Sigma t^j Y_j] = \sum_{i+j\leq k} t^{i+j}[X_i,Y_j]$. Study the Gelfand-Fuks cohomology of $V_{R^n}^{(k)}$.

This is related to deformations of foliations. A few variations follow:

6.1. Restrict the coefficients X_i to vector fields with compact support.

6.2. Replace the coefficients by formal vector fields on R^n.

6.3. Consider ℓ-parameter deformations, replacing $t\in R$ by $(t_1,\ldots,t_\ell)\in R^\ell$.

7. (Bott) Extend the proof of the Bott-Fuks conjecture (Haefliger, Segal, Bott, Stasheff-Anderson, and others) to complex manifolds, for C^∞ vector fields of type (1,0).

The diagonal complex C_Δ^* is believed to be known.

8. (Haefliger) Study the cohomology of continuous alternating forms on V_M with coefficients in a non-trivial representation.

Losik, Gelfand, and Fuks have initiated the study of this problem, and have calculated the diagonal complex C_{Δ}^{*} in the case that the coefficient algebra is the algebra $\Omega^{*}(M)$ of differental forms on M.

Ref. I.M. Gelfand and D.B. Fuks, Cohomology of the Lie algebra of vector fields with nontrivial coefficients, Functional Anal. Appl. 4(1970),181-192.

9. (Shulman) Study the Gelfand-Fuks cohomology of vector fields tangent to a foliation.

(This structure is probably so large that it is intractable.)

10. (Bott and Haefliger) Does S^6 admit a $B\Gamma_3^C$ structure whose normal bundle is given by the usual almost complex structure on S^6 ?

2. Variation of Foliations and Stability

11. Stability of foliations (Rosenberg). Let $\pi : E \to B$ be a smooth fibration with compact fiber L. Give sufficient conditions for every foliation F sufficiently close to the fibration to have a compact leaf.

In the C^1 case, $H^1(L;R) = 0$ is sufficient [Langevin and Rosenberg, On stability of compact leaves and fibrations, Topology 16 (1977),107-112]. For other positive results, see the two following articles.

Ref. R. Langevin and H. Rosenberg, Integrable perturbations of fibrations and a theorem of Seifert, these Proceedings.
J. Palis, Rigidity of the centralizers of diffeomorphisms and structural stability of suspended foliations, these Proceedings.

11.1. In particular, are the conditions $L = T^k$, π given by an action of R^k, and $\chi(B) \neq 0$ sufficient?

Yes, for a C^0 neighborhood of the Hopf fibration $S^3 \to S^2$ [H. Seifert, Closed integral curves in 3-space ..., Proc.A.M.S. 1 (1950),287-302] and, more generally, whenever $k = 1$ (F.B. Fuller, implicit in [An index of fixed-point type for periodic orbits, Amer.

J. Math. 89 (1967),133-148, esp. pp. 142-143], and explicit in [Fuller, The existence of periodic orbits, preprint, Univ. de Strasbourg, 1968]).

11.2. For the case $k = 2$, $B = S^2$, the question is equivalent to the following problem: Does every pair of commuting diffeomorphisms $f,g : S^2 \to S^2$ sufficiently C^1-close to the identity have a common fixed (or at least periodic) point?

The answer to this question is positive if f and g embed in commuting flows [E. Lima, Comment.Math.Helv. 39 (1964),97-110].

12. The space of foliations (Rosenberg). Is the space of C^r foliations of a fixed manifold M, with the C^r topology ($r \geq 1$) locally path-connected?

In the particular case that F is the product foliation of T^3 by tori, the question reduces (as in 11.2) to the following related problem: Given two commuting diffeomorphisms $f,g: S^1 \to S^1$ C^r-close to the identity, are they isotopic to the identity through commuting diffeomorphisms? It is known that if f and g are sufficiently C^2-close to the identity, then there exists a C^0 path of commuting homeomorphism pairs joining (f,g) to (id,id), such that the path stays close to (id,id) (Rosenberg).

For a discussion of the C^r topology of the space of C^r foliations, see [D.B.A. Epstein, A topology for the space of foliations, Geometry and Topology, Rio 1976, Springer LNM 597 (1977),132-150].

13. (Rosenberg) Let F be a smooth, C^1-stable foliation of S^3 with only generic (i.e., Morse) singularities. Are all leaves simply connected?

Ref. Rosenberg and Roussarie, J. Differential Geom. 10 (1975),p.219.

14. Reeb stability for non-compact leaves (Hector). Is there a stability theorem of Reeb-Thurston type for a proper leaf L of a C^2 codimension one foliation F of a compact manifold, if $H^1(L;R) = 0$?

If I is an open transverse arc such that the equivalence relation induced by F on I is trivial, then the saturated set of I is trivially foliated [G. Hector, Croissance des feuilletages presque sans holonomie, preprint, Section II]. Methods of J. Cantwell and L. Conlon [Poincaré-Bendixson theory for leaves of codimension one, preprint, Washington Univ. (St.Louis), 1977] give a sufficient condition for leaves of "finite class".

In the case of compact leaves, Langevin and Rosenberg have adapted Thurston's generalization of Reeb stability to the second stability theorem (stability under perturbation), showing that the condition $H^1(L;R) = 0$ is sufficient.

Ref. W. Thurston, A generalization of the Reeb stability theorem, Topology 13 (1974), 347-352.

R. Langevin and H. Rosenberg, On stability of compact leaves and fibrations, Topology 16 (1977), 107-111.

15. Study transversality for foliation singularities, in order to arrive at a suitable notion of genericity of foliation singularities.

The transversality lemma one would naturally seek is false [Thom, On singularities of foliations, Manifolds, Tokyo, 1973, p.172].

16. Cobordism of foliations (Rosenberg). Study cobordism of foliations. The foliation of the cobordism is required to be transverse to the boundary, so that a foliation of the same codimension is induced on the boundary.

Sergeraert has recently shown that every C^∞ Reeb foliation of S^3 bounds a compact foliated 4-manifold. For oriented 3-manifolds with codimension one foliations, the Godbillon-Vey invariant evaluated on the fundamental class is a cobordism invariant. Other characteristic classes of foliations give cobordism invariants in a similar way. Another kind of foliated cobordism, in which the foliation is tangent to the boundary of the cobordism, has been studied by G. Whiston [J. Differential Geometry 11 (1976), 475-478].

Ref. H. Rosenberg and W. Thurston, Some remarks on foliations, Dynamical Systems, Salvador, 1971, p. 478.

F. Sergeraert, Feuilletages et difféomorphismes infiniment tangents à l'identité, preprint, Poitiers, 1977.

R. Bott and A. Haefliger, On characteristic classes of Γ-foliations, Bull.A.M.S. 78 (1972), 1039-1044.

3. Qualitative Properties of Foliations

17. Holonomy and the Godbillon-Vey invariant (Rosenberg). If F is a C^∞ foliation without holonomy, is its Godbillon-Vey invariant gv(F)=0 [Rosenberg-Thurston, see preceding Problem]?

Yes, for a foliation of T^3 by planes [Herman, see Problem 3]. Here are some variants of this problem.

17.1. If F has no exceptional leaves, is gv(F) = 0 ?

If F has no holonomy and the underlying manifold is compact, then F has no exceptional leaves [R. Sacksteder, Foliations and pseudogroups, Amer.J.Math. 87 (1965), Theorems 1 and 6].

17.2. (Shulman) If F has no transversal holonomy [Godbillon, C.R.Acad. Sci.Paris 264 (1967) Série A, p. 1050], is gv(F) = 0 ?

Godbillon shows that there is no holonomy whenever there is no transversal holonomy.

17.3. (D.Sullivan) If all leaves of F have non-exponential growth, is gv(F) = 0 ?

When all leaves are non-compact with non-exponential growth and M is compact, then F has no holonomy [Plante, Ann.Inst.Fourier,Grenoble 25 (1975),p.248].

17.4. (D.Sullivan) More generally, what implications in the spirit of dynamical systems does the Godbillon-Vey invariant have?

Ref. (for some geometric insight into the Godbillon-Vey invariant):

B.Reinhart and J. Wood, A metric formula for the Godbillon-Vey invariant for foliations, Proc.A.M.S. 38(1973),427-430.

W. Thurston, Non-cobordant foliations of S^3, Bull. A.M.S. 78 (1972), 511-514.

18. Holonomy and exceptional leaves (Hector and Lamoureux). Can a C^∞ foliation with no holonomy have exceptional leaves?

If so, the underlying manifold M cannot be compact (Sacksteder, see Problem 17.1], $\pi_1(M)$ cannot be free abelian or finitely generated abelian, and rank $H_1(M;Q) > 1$ [Lamoureux, Dijon Colloquium 1974, Springer LNM 484, p. 267].

18.1. In particular, can there exist such a foliation transverse to the fibers of a foliation by circles?

If so, the fibration must be non-trivial (Lamoureux).

19. (Shulman) Let Ω be a closed non-singular q-form locally decomposable as a product of 1-forms, on a 4k-manifold M. (Thus Ω defines a codimension q foliation F with a holonomy-invariant transverse volume form.)

Prove: If q is odd then index(M) = 0 .

If F is a fibration (in particular, if $q = 1$ [Tischler, Topology 9(1970),p.153]),then index(M) = 0. On the other hand, Mizutani [Topology 13(1974),353-362] has constructed a foliated 4-manifold with non-zero index.

20. (Sullivan) Let F_1 and F_2 be transversal smooth foliations of a connected manifold M with complementary dimensions k and ℓ such that all leaves are planes with non-exponential growth.

Conjecture. The universal cover $\tilde{M}$ is diffeomorphic to $R^k \times R^\ell$ such that F_1 and F_2 lift to the product foliations.

When exponential growth is allowed, there is a counterexample (Deligne; perhaps known earlier). Let $G = SL(3,R)$ and define

$$H_1 = \left\{\begin{pmatrix} a & b & c \\ & a^{-1} & d \\ & & 1 \end{pmatrix}\right\} \text{ and } H_2 = \left\{\begin{pmatrix} 1 & & \\ b & a & \\ c & d & a^{-1} \end{pmatrix}\right\}, \quad a,b,c,d \in R, a>0.$$

Let Γ be a discrete subgroup of G meeting each subgroup conjugate to H_1 or H_2 only in the identity I, and such that G/Γ is compact. Then the foliations F_1 (resp. F_2) of G/Γ with leaves $H_1 g\Gamma$ (resp. $H_2 g\Gamma$) for each $g \in G$, are transversal, and all their leaves are 4-planes, but $\tilde{M} = S^3 \times R^5$.

21. (Lamoureux) A C^2 codimension one foliation without holonomy on a compact manifold has a trivially foliated universal cover [Sacksteder, see Problem 17.1, Theorem 6]. Does this hold for C^1 foliations? Can

one find a direct, geometric proof?

By Reeb stability, the conclusion holds for a C^0 foliation with a compact leaf.

22. Trapped leaves (Lamoureux). Must every leaf of a C^∞ codimension one foliation F, cut by a closed, null-homologous transversal τ, be trapped?

A leaf is trapped ("captée") iff there is non-trivial holonomy in its closure.

If F has no holonomy, then M must be non-compact [Sacksteder, see Problem 17.7] and every leaf cut by τ must be locally dense [Lamoureux, Ann.Inst.Fourier,Grenoble 26(1976), Theorem B, p. 232]. If M is compact, then the answer is affirmative ([Sacksteder] and L.Conlon), provided M and F are orientable."

23. Vanishing cycles and holonomy (Lamoureux). Give a direct proof that a codimension one foliation of a compact manifold with a vanishing cycle has holonomy.

This follows from Theorem 6 of [Sacksteder, see Problem 17.7] if the foliation is C^2.

24. Novikov's Theorem for C^0 foliations. Prove that every C^0 codimension one foliation of S^3 contains a compact leaf (and, in fact, a Reeb component).

Ref. S.P. Novikov, Trans. Moscow Math.Soc. 14 (1965), 268-304.
A. Haefliger, Sem. Bourbaki (Feb. 1968) Nº 339.

25. Hilbert's 16th Problem. Show that the number of closed orbits of a vector field on R^2 given by two polynomials of degree at most n is bounded by a number depending only on n.

Ref. C.Pugh, Hilbert's 16th problem, Dynamical Systems--Warwick 1974, Springer LNM 468(1975), 55-57.

26. (Reeb) Let $\omega = dz + P(x,y,z)dx + Q(x,y,z)dy$ be an integrable 1-form with polynomial coefficients on R^3. Study the relationship bet-

ween the polynomials P and Q and the leaf space B.

Since ω is analytic, the leaves are closed and project diffeomorphically onto open sets of R^2. Consequently, B is a (possibly non-Hausdorff) 1-manifold [Haefliger, Comment.Math.Helv. 32 (1958), p. 387].

27. Extension of differential forms (Reeb). Define the extension r of a Pfaffian form $\omega \in \Lambda^1(M)$ to be the smallest integer such that there exist $f_1,\ldots,f_r \in C^\infty(M;R)$ with either

$$\omega = df_1 + f_2 df_3 + \ldots + f_{r-1} df_r \qquad (r \text{ odd}), \text{ or}$$

$$\omega = f_1 df_2 + \ldots + f_{r-1} df_r \qquad (r \text{ even}).$$

Problem. Study $r(\omega)$, in particular when ω is integrable.

For a form of constant class, Darboux's theorem says that locally the extension coincides with the class. Every ω can be globally C^0-approximated by a Pfaffian form $\omega_1 = df_1 + f_2 df_3$ [Varela, Ann. Inst.Fourier,Grenoble 26 (1976), 239-271].

28. Exceptional minimal sets (Hector). Let C be an exceptional minimal set of a C^2 codimension one foliation F of a compact manifold M. Show

1. M-C has a finite number of connected components (false if F is only assumed to be C^1 (Hector).

2. C contains a leaf whose fundamental group is not finitely generated.

29. (Hector) Can a C^2 codimension one foliation of a compact manifold have all its leaves exceptional?

Yes, if F is only C^0. Furthermore, there exists a C^∞ foliation of R^3 with all leaves exceptional [Hector, Ann.Inst. Fourier, Grenoble 26(1976), p. 256].

4. Minimal Sets

30. (Rosenberg) What 3-manifolds can be minimal sets of smooth flows?

A manifold is a minimal set of a flow (or homeomorphism) if every orbit is dense. In this case the flow is also said to be minimal. For example, the horocycle flow on $SL(2,\mathbb{R})/\Gamma$ is minimal if Γ is a uniform discrete subgroup.

30.1. The Gottschalk Conjecture: There is no minimal flow on S^3.

There does exist a C^∞ flow on S^3 with almost every orbit dense, since every compact connected C^∞ manifold of dimension $n \geq 3$ has a C^∞ flow that is ergodic with respect to a smooth measure [Anosov, Math.U.S.S.R.Izv. 8 (1974) Nº 3, 525-552]. There also exist a C^∞ minimal diffeomorphism on S^3, and C^∞ minimal flows on any manifold which admits a free action of the torus T^2 [A. Fathi and M.R. Herman, Existence de difféomorphismes minimaux, to appear].

31. The Seifert Conjecture. Does every C^2 (or C^∞) flow on S^3 have a compact orbit?

The answer is no for C^1 flows [Schweitzer, Annals of Math. 100 (1974), 386-400]. A positive answer would imply the Gottschalk conjecture (preceding problem).

32. (Schweitzer) Does every C^2 (or C^∞) codimension one foliation of S^5 have a compact leaf?

Every smooth manifold of dimension ≥ 5 with $\chi = 0$ has a C^0 codimension one foliation with no compact leaf [B. Raymond, Ensembles de Cantor et feuilletages, preprint, Orsay, 1976, and Schweitzer, Proc.A.M.S.Symp.Pure Math. 27(1) (1975), 311-312].

33. (M.Herman) Does $S^1 \times R$ have a homeomorphism for which it is minimal?

33.1. More generally, what manifolds admit minimal homeomorphisms?

There is no minimal homeomorphism on R^2, because for any orientation preserving homeomorphism f of R^2 without fixed points there exists a disk $D \subset R^2$ which does not meet its translate $f^n(D)$ for any integer $n \neq 0$ [L.E.J. Brouwer, Math.Annalen

72 (1912), 36-54]. On the other hand, it is easy to construct a minimal flow on $T^2 - \{(0,0)\}$, by simply multiplying a constant, irrational slope vector field by a non-negative function which vanishes exactly at the missing point (0,0) [Nemytskii and Stepanov, Qualitative Theory of Differential Equations, Ex. 4.06, p. 346].

Any compact connected manifold with a locally free action of S^1 has a C^∞ minimal diffeomorphism [Fathi and Herman, see Problem 30.1].

34. Does every foliation of R^3 by curves have a minimal set?

34.1. Does there exist any manifold M^n with a flow that has no minimal set?

Of course, M^n would have to be non-compact.

35. Can R^3 be a minimal set of a foliation by curves?

36. (Epstein) Can R^3 be foliated by circles?

It is possible to decompose R^3 as a union of pairwise disjoint smooth circles, but in the known examples the circles do not form a foliation (L. Markus).

Vol. 489: J. Bair and R. Fourneau, Etude Géométrique des Espaces Vectoriels. Une Introduction. VII, 185 pages. 1975.

Vol. 490: The Geometry of Metric and Linear Spaces. Proceedings 1974. Edited by L. M. Kelly. X, 244 pages. 1975.

Vol. 491: K. A. Broughan, Invariants for Real-Generated Uniform Topological and Algebraic Categories. X, 197 pages. 1975.

Vol. 492: Infinitary Logic: In Memoriam Carol Karp. Edited by D. W. Kueker. VI, 206 pages. 1975.

Vol. 493: F. W. Kamber and P. Tondeur, Foliated Bundles and Characteristic Classes. XIII, 208 pages. 1975.

Vol. 494: A Cornea and G. Licea. Order and Potential Resolvent Families of Kernels. IV, 154 pages. 1975.

Vol. 495: A. Kerber, Representations of Permutation Groups II. V, 175 pages.1975.

Vol. 496: L. H. Hodgkin and V. P. Snaith, Topics in K-Theory. Two Independent Contributions. III, 294 pages. 1975.

Vol. 497: Analyse Harmonique sur les Groupes de Lie. Proceedings 1973–75. Edité par P. Eymard et al. VI, 710 pages. 1975.

Vol. 498: Model Theory and Algebra. A Memorial Tribute to Abraham Robinson. Edited by D. H. Saracino and V. B. Weispfenning. X, 463 pages. 1975.

Vol. 499: Logic Conference, Kiel 1974. Proceedings. Edited by G. H. Müller, A. Oberschelp, and K. Potthoff. V, 651 pages 1975.

Vol. 500: Proof Theory Symposion, Kiel 1974. Proceedings. Edited by J. Diller and G. H. Müller. VIII, 383 pages. 1975.

Vol. 501: Spline Functions, Karlsruhe 1975. Proceedings. Edited by K. Böhmer, G. Meinardus, and W. Schempp. VI, 421 pages. 1976.

Vol. 502: János Galambos, Representations of Real Numbers by Infinite Series. VI, 146 pages. 1976.

Vol. 503: Applications of Methods of Functional Analysis to Problems in Mechanics. Proceedings 1975. Edited by P. Germain and B. Nayroles. XIX, 531 pages. 1976.

Vol. 504: S. Lang and H. F. Trotter, Frobenius Distributions in GL_2-Extensions. III, 274 pages. 1976.

Vol. 505: Advances in Complex Function Theory. Proceedings 1973/74. Edited by W. E. Kirwan and L. Zalcman. VIII, 203 pages. 1976.

Vol. 506: Numerical Analysis, Dundee 1975. Proceedings. Edited by G. A. Watson. X, 201 pages. 1976.

Vol. 507: M. C. Reed, Abstract Non-Linear Wave Equations. VI, 128 pages. 1976.

Vol. 508: E. Seneta, Regularly Varying Functions. V, 112 pages. 1976.

Vol. 509: D. E. Blair, Contact Manifolds in Riemannian Geometry. VI, 146 pages. 1976.

Vol. 510: V. Poènaru, Singularités C^∞ en Présence de Symétrie. V, 174 pages. 1976.

Vol. 511: Séminaire de Probabilités X. Proceedings 1974/75. Edité par P. A. Meyer. VI, 593 pages. 1976.

Vol. 512: Spaces of Analytic Functions, Kristiansand, Norway 1975. Proceedings. Edited by O. B. Bekken, B. K. Øksendal, and A. Stray. VIII, 204 pages. 1976.

Vol. 513: R. B. Warfield, Jr. Nilpotent Groups. VIII, 115 pages. 1976.

Vol. 514: Séminaire Bourbaki vol. 1974/75. Exposés 453 – 470. IV, 276 pages. 1976.

Vol. 515: Bäcklund Transformations. Nashville, Tennessee 1974. Proceedings. Edited by R. M. Miura. VIII, 295 pages. 1976.

Vol. 516: M. L. Silverstein, Boundary Theory for Symmetric Markov Processes. XVI, 314 pages. 1976.

Vol. 517: S. Glasner, Proximal Flows. VIII, 153 pages. 1976.

Vol. 518: Séminaire de Théorie du Potentiel, Proceedings Paris 1972–1974. Edité par F. Hirsch et G. Mokobodzki. VI, 275 pages. 1976.

Vol. 519: J. Schmets, Espaces de Fonctions Continues. XII, 150 pages. 1976.

Vol. 520: R. H. Farrell, Techniques of Multivariate Calculation. X, 337 pages. 1976.

Vol. 521: G. Cherlin, Model Theoretic Algebra – Selected Topics. IV, 234 pages. 1976.

Vol. 522: C. O. Bloom and N. D. Kazarinoff, Short Wave Radiation Problems in Inhomogeneous Media: Asymptotic Solutions. V. 104 pages. 1976.

Vol. 523: S. A. Albeverio and R. J. Høegh-Krohn, Mathematical Theory of Feynman Path Integrals. IV, 139 pages. 1976.

Vol. 524: Séminaire Pierre Lelong (Analyse) Année 1974/75. Edité par P. Lelong. V, 222 pages. 1976.

Vol. 525: Structural Stability, the Theory of Catastrophes, and Applications in the Sciences. Proceedings 1975. Edited by P. Hilton. VI, 408 pages. 1976.

Vol. 526: Probability in Banach Spaces. Proceedings 1975. Edited by A. Beck. VI, 290 pages. 1976.

Vol. 527: M. Denker, Ch. Grillenberger, and K. Sigmund, Ergodic Theory on Compact Spaces. IV, 360 pages. 1976.

Vol. 528: J. E. Humphreys, Ordinary and Modular Representations of Chevalley Groups. III, 127 pages. 1976.

Vol. 529: J. Grandell, Doubly Stochastic Poisson Processes. X, 234 pages. 1976.

Vol. 530: S. S. Gelbart, Weil's Representation and the Spectrum of the Metaplectic Group. VII, 140 pages. 1976.

Vol. 531: Y.-C. Wong, The Topology of Uniform Convergence on Order-Bounded Sets. VI, 163 pages. 1976.

Vol. 532: Théorie Ergodique. Proceedings 1973/1974. Edité par J.-P. Conze and M. S. Keane. VIII, 227 pages. 1976.

Vol. 533: F. R. Cohen, T. J. Lada, and J. P. May, The Homology of Iterated Loop Spaces. IX, 490 pages. 1976.

Vol. 534: C. Preston, Random Fields. V, 200 pages. 1976.

Vol. 535: Singularités d'Applications Differentiables. Plans-sur-Bex. 1975. Edité par O. Burlet et F. Ronga. V, 253 pages. 1976.

Vol. 536: W. M. Schmidt, Equations over Finite Fields. An Elementary Approach. IX, 267 pages. 1976.

Vol. 537: Set Theory and Hierarchy Theory. Bierutowice, Poland 1975. A Memorial Tribute to Andrzej Mostowski. Edited by W. Marek, M. Srebrny and A. Zarach. XIII, 345 pages. 1976.

Vol. 538: G. Fischer, Complex Analytic Geometry. VII, 201 pages. 1976.

Vol. 539: A. Badrikian, J. F. C. Kingman et J. Kuelbs, Ecole d'Eté de Probabilités de Saint Flour V-1975. Edité par P.-L. Hennequin. IX, 314 pages. 1976.

Vol. 540: Categorical Topology, Proceedings 1975. Edited by E. Binz and H. Herrlich. XV, 719 pages. 1976.

Vol. 541: Measure Theory, Oberwolfach 1975. Proceedings. Edited by A. Bellow and D. Kölzow. XIV, 430 pages. 1976.

Vol. 542: D. A. Edwards and H. M. Hastings, Čech and Steenrod Homotopy Theories with Applications to Geometric Topology. VII, 296 pages. 1976.

Vol. 543: Nonlinear Operators and the Calculus of Variations, Bruxelles 1975. Edited by J. P. Gossez, E. J. Lami Dozo, J. Mawhin, and L. Waelbroeck, VII, 237 pages. 1976.

Vol. 544: Robert P. Langlands, On the Functional Equations Satisfied by Eisenstein Series. VII, 337 pages. 1976.

Vol. 545: Noncommutative Ring Theory. Kent State 1975. Edited by J. H. Cozzens and F. L. Sandomierski. V, 212 pages. 1976.

Vol. 546: K. Mahler, Lectures on Transcendental Numbers. Edited and Completed by B. Diviš and W. J. Le Veque. XXI, 254 pages. 1976.

Vol. 547: A. Mukherjea and N. A. Tserpes, Measures on Topological Semigroups: Convolution Products and Random Walks. V, 197 pages. 1976.

Vol. 548: D. A. Hejhal, The Selberg Trace Formula for PSL (2,$\mathbb{R}$). Volume I. VI, 516 pages. 1976.

Vol. 549: Brauer Groups, Evanston 1975. Proceedings. Edited by D. Zelinsky. V, 187 pages. 1976.

Vol. 550: Proceedings of the Third Japan – USSR Symposium on Probability Theory. Edited by G. Maruyama and J. V. Prokhorov. VI, 722 pages. 1976.

Vol. 551: Algebraic K-Theory, Evanston 1976. Proceedings. Edited by M. R. Stein. XI, 409 pages. 1976.

Vol. 552: C. G. Gibson, K. Wirthmüller, A. A. du Plessis and E. J. N. Looijenga. Topological Stability of Smooth Mappings. V, 155 pages. 1976.

Vol. 553: M. Petrich, Categories of Algebraic Systems. Vector and Projective Spaces, Semigroups, Rings and Lattices. VIII, 217 pages. 1976.

Vol. 554: J. D. H. Smith, Mal'cev Varieties. VIII, 158 pages. 1976.

Vol. 555: M. Ishida, The Genus Fields of Algebraic Number Fields. VII, 116 pages. 1976.

Vol. 556: Approximation Theory. Bonn 1976. Proceedings. Edited by R. Schaback and K. Scherer. VII, 466 pages. 1976.

Vol. 557: W. Iberkleid and T. Petrie, Smooth S^1 Manifolds. III, 163 pages. 1976.

Vol. 558: B. Weisfeiler, On Construction and Identification of Graphs. XIV, 237 pages. 1976.

Vol. 559: J.-P. Caubet, Le Mouvement Brownien Relativiste. IX, 212 pages. 1976.

Vol. 560: Combinatorial Mathematics, IV, Proceedings 1975. Edited by L. R. A. Casse and W. D. Wallis. VII, 249 pages. 1976.

Vol. 561: Function Theoretic Methods for Partial Differential Equations. Darmstadt 1976. Proceedings. Edited by V. E. Meister, N. Weck and W. L. Wendland. XVIII, 520 pages. 1976.

Vol. 562: R. W. Goodman, Nilpotent Lie Groups: Structure and Applications to Analysis. X, 210 pages. 1976.

Vol. 563: Séminaire de Théorie du Potentiel. Paris, No. 2. Proceedings 1975–1976. Edited by F. Hirsch and G. Mokobodzki. VI, 292 pages. 1976.

Vol. 564: Ordinary and Partial Differential Equations, Dundee 1976. Proceedings. Edited by W. N. Everitt and B. D. Sleeman. XVIII, 551 pages. 1976.

Vol. 565: Turbulence and Navier Stokes Equations. Proceedings 1975. Edited by R. Temam. IX, 194 pages. 1976.

Vol. 566: Empirical Distributions and Processes. Oberwolfach 1976. Proceedings. Edited by P. Gaenssler and P. Révész. VII, 146 pages. 1976.

Vol. 567: Séminaire Bourbaki vol. 1975/76. Exposés 471–488. IV, 303 pages. 1977.

Vol. 568: R. E. Gaines and J. L. Mawhin, Coincidence Degree, and Nonlinear Differential Equations. V, 262 pages. 1977.

Vol. 569: Cohomologie Etale SGA $4\frac{1}{2}$. Séminaire de Géométrie Algébrique du Bois-Marie. Edité par P. Deligne. V, 312 pages. 1977.

Vol. 570: Differential Geometrical Methods in Mathematical Physics, Bonn 1975. Proceedings. Edited by K. Bleuler and A. Reetz. VIII, 576 pages. 1977.

Vol. 571: Constructive Theory of Functions of Several Variables, Oberwolfach 1976. Proceedings. Edited by W. Schempp and K. Zeller. VI, 290 pages. 1977

Vol. 572: Sparse Matrix Techniques, Copenhagen 1976. Edited by V. A. Barker. V, 184 pages. 1977.

Vol. 573: Group Theory, Canberra 1975. Proceedings. Edited by R. A. Bryce, J. Cossey and M. F. Newman. VII, 146 pages. 1977.

Vol. 574: J. Moldestad, Computations in Higher Types. IV, 203 pages. 1977.

Vol. 575: K-Theory and Operator Algebras, Athens, Georgia 1975. Edited by B. B. Morrel and I. M. Singer. VI, 191 pages. 1977.

Vol. 576: V. S. Varadarajan, Harmonic Analysis on Real Reductive Groups. VI, 521 pages. 1977.

Vol. 577: J. P. May, E_∞ Ring Spaces and E_∞ Ring Spectra. IV, 268 pages. 1977.

Vol. 578: Séminaire Pierre Lelong (Analyse) Année 1975/76. Edité par P. Lelong. VI, 327 pages. 1977.

Vol. 579: Combinatoire et Représentation du Groupe Symétrique, Strasbourg 1976. Proceedings 1976. Edité par D. Foata. IV, 339 pages. 1977.

Vol. 580: C. Castaing and M. Valadier, Convex Analysis and Measurable Multifunctions. VIII, 278 pages. 1977.

Vol. 581: Séminaire de Probabilités XI, Université de Strasbourg. Proceedings 1975/1976. Edité par C. Dellacherie, P. A. Meyer et M. Weil. VI, 574 pages. 1977.

Vol. 582: J. M. G. Fell, Induced Representations and Banach *-Algebraic Bundles. IV, 349 pages. 1977.

Vol. 583: W. Hirsch, C. C. Pugh and M. Shub, Invariant Manifolds. IV, 149 pages. 1977.

Vol. 584: C. Brezinski, Accélération de la Convergence en Analyse Numérique. IV, 313 pages. 1977.

Vol. 585: T. A. Springer, Invariant Theory. VI, 112 pages. 1977.

Vol. 586: Séminaire d'Algèbre Paul Dubreil, Paris 1975–1976 (29ème Année). Edited by M. P. Malliavin. VI, 188 pages. 1977.

Vol. 587: Non-Commutative Harmonic Analysis. Proceedings 1976. Edited by J. Carmona and M. Vergne. IV, 240 pages. 1977.

Vol. 588: P. Molino, Théorie des G-Structures: Le Problème d'Equivalence. VI, 163 pages. 1977.

Vol. 589: Cohomologie l-adique et Fonctions L. Séminaire de Géométrie Algébrique du Bois-Marie 1965–66, SGA 5. Edité par L. Illusie. XII, 484 pages. 1977.

Vol. 590: H. Matsumoto, Analyse Harmonique dans les Systèmes de Tits Bornologiques de Type Affine. IV, 219 pages. 1977.

Vol. 591: G. A. Anderson, Surgery with Coefficients. VIII, 157 pages. 1977.

Vol. 592: D. Voigt, Induzierte Darstellungen in der Theorie der endlichen, algebraischen Gruppen. V, 413 Seiten. 1977.

Vol. 593: K. Barbey and H. König, Abstract Analytic Function Theory and Hardy Algebras. VIII, 260 pages. 1977.

Vol. 594: Singular Perturbations and Boundary Layer Theory, Lyon 1976. Edited by C. M. Brauner, B. Gay, and J. Mathieu. VIII, 539 pages. 1977.

Vol. 595: W. Hazod, Stetige Faltungshalbgruppen von Wahrscheinlichkeitsmaßen und erzeugende Distributionen. XIII, 157 Seiten. 1977.

Vol. 596: K. Deimling, Ordinary Differential Equations in Banach Spaces. VI, 137 pages. 1977.

Vol. 597: Geometry and Topology, Rio de Janeiro, July 1976. Proceedings. Edited by J. Palis and M. do Carmo. VI, 866 pages. 1977.

Vol. 598: J. Hoffmann-Jørgensen, T. M. Liggett et J. Neveu, Ecole d'Eté de Probabilités de Saint-Flour VI – 1976. Edité par P.-L. Hennequin. XII, 447 pages. 1977.

Vol. 599: Complex Analysis, Kentucky 1976. Proceedings. Edited by J. D. Buckholtz and T. J. Suffridge. X, 159 pages. 1977.

Vol. 600: W. Stoll, Value Distribution on Parabolic Spaces. VIII, 216 pages. 1977.

Vol. 601: Modular Functions of one Variable V, Bonn 1976. Proceedings. Edited by J.-P. Serre and D. B. Zagier. VI, 294 pages. 1977.

Vol. 602: J. P. Brezin, Harmonic Analysis on Compact Solvmanifolds. VIII, 179 pages. 1977.

Vol. 603: B. Moishezon, Complex Surfaces and Connected Sums of Complex Projective Planes. IV, 234 pages. 1977.

Vol. 604: Banach Spaces of Analytic Functions, Kent, Ohio 1976. Proceedings. Edited by J. Baker, C. Cleaver and Joseph Diestel. VI, 141 pages. 1977.

Vol. 605: Sario et al., Classification Theory of Riemannian Manifolds. XX, 498 pages. 1977.

Vol. 606: Mathematical Aspects of Finite Element Methods. Proceedings 1975. Edited by I. Galligani and E. Magenes. VI, 362 pages. 1977.

Vol. 607: M. Métivier, Reelle und Vektorwertige Quasimartingale und die Theorie der Stochastischen Integration. X, 310 Seiten. 1977.

Vol. 608: Bigard et al., Groupes et Anneaux Réticulés. XIV, 334 pages. 1977.

MIX
Papier aus verantwortungsvollen Quellen
Paper from responsible sources
FSC® C105338
FSC
www.fsc.org

If you have any concerns about our products,
you can contact us on
ProductSafety@springernature.com

In case Publisher is established outside the EU,
the EU authorized representative is:
Springer Nature Customer Service Center GmbH
Europaplatz 3, 69115 Heidelberg, Germany

Printed by Libri Plureos GmbH
in Hamburg, Germany